ICMI Study Series Editors A.G. Howson and J.-P. Kahane

Mathematics and Cognition:
A Research Synthesis by the International Group
for the Psychology of Mathematics Education

Edited by
Pearla Nesher and Jeremy Kilpatrick

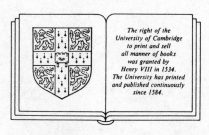

*The right of the
University of Cambridge
to print and sell
all manner of books
was granted by
Henry VIII in 1534.
The University has printed
and published continuously
since 1584.*

CAMBRIDGE UNIVERSITY PRESS
Cambridge
New York Port Chester
Melbourne Sydney

Published by the Press Syndicate of the University of Cambridge
The Pitt Building, Trumpington Street, Cambridge CB2 1RP
40 West 20th Street, New York, NY 10011, USA
10 Stamford Road, Oakleigh, Melbourne, 3166, Australia

First Published 1990

Printed in Great Britain at the University Press, Cambridge

Library of Congress cataloging in publication data available

British Library Cataloguing in publication data available

ISBN 0 521 36608 9 Hardcovers
ISBN 0 521 36787 5 Paperback

Contents

Foreword

This book is aimed at presenting to a wider readership the major findings of research related to mathematics education carried out over the last 10 years by the International Group for the Psychology of Mathematics Education (PME). This group was organized in 1976 at the Third International Congress on Mathematical Education in Karlsruhe, West Germany, but as Efraim Fischbein, the organizer, notes in the introduction, its roots go back to the First ICME in Lyons, France, in 1969. PME is affiliated with the International Commission for Mathematics Instruction (ICMI) and has about 500 members from 39 countries.

The purpose of the book is to present major findings of research in mathematics education that relate to the cognitive aspects of teaching and learning. The audience consists of teachers, whether in service or preparing to teach; teacher educators; educational researchers; cognitive psychologists; and the general reader who is interested in mathematics education from a psychological point of view. Each chapter reviews recent research related to a single theme. Most of the research cited has been reported in sessions at the annual meetings of PME or its North American branch, but studies from other sources are included to round out the reviews.

It should be emphasized that, as befits a field of inquiry that is still being formed, this book is truly a collaboration. Although one or two people took the principal responsibility for writing the final draft of the chapter, each chapter was originally constructed by a group of PME members active in research related to the theme. The themes and contributors were selected at the time of the PME meeting in Montréal, Canada, in 1987; draft chapters were critiqued in a conference of principal contributors after the PME meeting in Veszprém, Hungary, in 1988, and then final drafts were prepared. The names of the other contributors to each chapter are listed on the first page of the chapter.

The psychology of mathematics education is a relatively young field. It attempts to study the teaching and learning of mathematics in new ways. It does not deal with the logic of mathematics as a school subject; rather, it attempts to understand what learners face as they encounter mathematics. It assumes that the learning of mathematics has its own psychology, that students and teachers bring their own ideas about mathematics to any learning situation, and that teachers will be better equipped to teach mathematics if they can understand how the subject looks from the perspective of their students.

As the didactics of mathematics becomes more extensively informed by research of the sort reviewed in this book, it can contribute in deeper ways to improving teaching. The didactics of mathematics is not a collection of tricks. It is an arena in which cognitive research studies are being used to build islands of organized knowledge that can undergird and enlighten practice. Whether the studies are conducted in the classrooms of France or in the streets of Brazil, they can help a teacher in any country see the possible interpretations that students might make of a mathematical activity.

Seven themes were chosen for treatment in this book. Of necessity, therefore, many important topics are missing or underrepresented. For example, there is no discussion of the mathematics of the middle school years, such as multiplication and rational numbers. Problem solving is not treated as a theme--although it is mentioned in several of the chapters. There is no chapter, and little attention overall, devoted to the teaching process. These topics are not missing in the work of PME, but some things had to be left for consideration in the future.

After the introduction, the book begins with a discussion of the nature of mathematical knowledge. Then there are five chapters that consider aspects of mathematics learning in an order that corresponds roughly to the increasing age of the students in the studies reviewed, from the learning of counting and simple addition to the learning of functions and proof. The final chapter looks at directions in which research on mathematics and cognition might move in the future.

We thank the ICMI and PME for the financial assistance that allowed the principal contributors to confer in Veszprém and the editors to communicate expeditiously with the contributors.

Pearla Nesher
University of Haifa

Jeremy Kilpatrick
University of Georgia

Principal Contributors

Nicolas Balacheff
Université Joseph Fourier, Grenoble 1
Équipe de Didactique des Mathématiques
et de l'Informatique
Labo LSD-IMAG, B.P. 53 X
38041 Grenoble Cedex, France

Jacques C. Bergeron
Université de Montréal
Département de Didactique
C.P. 6128, Succ. A
Montréal, PQ   H3C 3J7, Canada

Tommy Dreyfus
Center for Technological Education Holon
P.O. Box 305
Holon 58102, Israel

Efraim Fischbein
Tel Aviv University
School of Education
Tel Aviv 69978, Israel

Nicolas Herscovics
Concordia University
7141 Sherbrooke W.
Montréal, PQ   H4B 1R6, Canada

Rina Hershkowitz
Weizmann Institute of Science
Department of Science Teaching
P.O. Box 26
Rehovot 76100, Israel

Carolyn Kieran
Université du Québec à Montréal
Département de Mathématiques et d'Informatique
C.P. 8888, Succ. A
Montréal, PQ   H3C 3P8, Canada

Colette Laborde
Université Joseph Fourier, Grenoble 1
Équipe de Didactique des Mathématiques
et de l'Informatique
Labo LSD-IMAG, B.P. 53 X
38041 Grenoble Cedex, France

Gérard Vergnaud
CNRS, GR Didactique
46, rue Saint-Jacques
75005 Paris, France

INTRODUCTION

Efraim Fischbein

Psychology and Mathematics

There has always been a reciprocal interest between psychology and mathematics despite the fact that these two domains are structurally so different. Psychologists, following the example of other empirical sciences, have tried to express psychological phenomena in terms of mathematical models, quantitative laws, and even axiomatic deductive systems (such as that of Clark Hull). On the other hand, mathematicians like Poincaré, Hadamard, and, more recently, Polya and Freudenthal became interested in the psychology of mathematical reasoning. Their accounts based on introspective attempts or didactical experiences are genuinely illuminating for the psychological aspects of mathematical reasoning. Piaget and his co-workers have played a fundamental role in developing the psychology of mathematical reasoning as a research domain. Their works on concepts like classes and relations, number, geometrical representations, and proportional and probabilistic reasoning have had an essential role in deepening our understanding of mathematical notions.

Nevertheless, mathematics education took little advantage of these developments. For a very long time, mathematical curricula and teaching methods were inspired mainly by new mathematical ideas like those promoted by the Bourbaki group or by the current experience of schoolteachers. Psychology and mathematics education as research domains remained relatively disjoint.

The split persisted primarily because of a lack of communication and consequently a lack of cooperation. A few basic mathematical concepts attracted the interest of psychologists, but most of the major concepts of mathematics remained outside their concern. The basic reason for that situation was a lack of professional competence by psychologists in the domain of mathematics. But it was also the implicit belief of mathematicians interested in education that a good command of mathematics, combined with natural didactical intuition, is the only requirement for teaching mathematics, defining curriculum content, creating teaching methods, and writing good textbooks. Most mathematicians did not consider seriously the idea that psycho-didactical research might be helpful for improving mathematics education. A mathematician who would never accept a theorem without a proof would be ready to put forward and strongly support educational suggestions in mathematics without relying on any explicit research evidence and without feeling the need to evaluate objectively and systematically the educational effects of his or her ideas.

In fact, mathematics education did not represent a defined research domain possessing its own theoretical background and a proper, adequate system of research strategies. Most of the activities promoted by the International Commission on Mathematics Instruction (ICMI) after its founding in 1909 consisted of organizing international comparative surveys on mathematics education and preparing international meetings devoted to curricula problems. The idea of basing educational reforms on scientific grounds and, more specifically, on psychological data was seldom expressed. As a matter of fact, it was Hans Freudenthal who, during a meeting organized by ICMI in Geneva in 1955, insisted on the necessity of promoting scientific research, including psychological considerations, to respond to problems raised by mathematics education (see Howson, 1984, p. 82).

As I have already said, it was primarily a lack of communication between mathematicians, psychologists, and mathematics teachers that prevented the creation of a genuine interdisciplinary domain. Psychology and mathematics are fundamentally different, not only in content but first and foremost in the nature of scientific reasoning. Mathematics is essentially a strictly formal deductive domain. The organization of a mathematical domain is typically axiomatic. A formal proof has a purely logical, deductive nature. Mathematical entities are never concrete objects. Every concept should be defined in an explicit, absolutely univocal manner.

In contrast, psychology belongs to the empirical sciences. The concepts used, mostly derived from practical or even introspective experiences, lack the exactitude, the precision, the univocal terminology of mathematical concepts. Psychological investigation has to take into account a large variety of subjective, social, and physical factors. The results obtained are of a statistical nature. One deals with masses of phenomena and not with what has been called *generic examples* (Mason & Pimm, 1984). If you want to prove the Pythagorean theorem, you consider one rectangular triangle and through this particular example you see the general properties. Your proof is definitive. The proved theorem is of universal validity. The exclusion of nonrelevant variables is made automatically by a trained mind, without any experimental intervention.

In psychological research, things are totally different. The final validity of the findings can never be established by purely logical considerations. The object of your research always contains much more potential information than that which can be expressed by definitions. To define variables in the psychological field is a very complex task. In fact, absolute distinctions can never be made. Terms like motivation, intent, and purpose overlap. Solving a problem is, in principle, a question of strategies of reasoning, but success depends so much on experience, knowledge, intuition, attention, emotion, motivation, and so forth, that a theoretical model that considers only the logical steps will be insufficient for predicting the solver's success.

A mathematician can produce very interesting, pertinent descriptions of his or her own mental processes. But this is not yet research if one considers the term research with its proper meaning. Asked to perform empirical-psychological research--with its qualities of objectivity and generality--the mathematician is usually perplexed.

The two domains, psychology and mathematics, as research domains, seem at first glance to be sharply different, which explains why it took so long before the psychology of mathematical activity became a domain of scientific investigation.

Researchers in mathematics education face a difficult problem. They should possess a clear, genuine, absolutely correct understanding of the mathematical concepts and explanations they are dealing with, and *at the same time* their ways of reasoning should be adapted to all the requirements of empirical, behavioral research. This task is clearly very complex. Despite a reciprocal interest, mathematics and psychology could blend only with difficulty into an interdisciplinary research domain.

An important impulse came from mathematicians interested in pedagogical problems. George Polya played an essential role in this direction. It is especially when one teaches mathematics that one discovers the deep similarity between endeavors in empirical and mathematical investigation, despite the sharp distinction between the two conceptual systems. In mathematics, as in empirical fields, when coping with a problem, one hypothesizes, one guesses, one tries, one experiments mentally, one learns from inductive findings, from analogies, and so forth. It is only the final product, the axiomatic organization of a mathematical body of knowledge, that is profoundly different from the usual organization of an empirical science. An awareness of these similarities contributed to the emergence of the psychology of mathematical activity as an interdisciplinary domain. When psychologists and mathematicians became aware of these common aspects, the conditions were ripe for creating the organizational framework for genuine cooperation.

The Creation of PME

At the first International Congress on Mathematical Education held in Lyons, France, in 1969, E. G. Begle, in his lecture, affirmed:

I see little hope for any further substantial improvements in mathematics education until we turn mathematics education into an experimental science, until we abandon our reliance on philosophical discussion based on dubious assumptions and instead follow a carefully correlated pattern of observation and speculation, the pattern so successfully employed by the physical and natural sciences. . . . To slight either the empirical observations or the theory building would be folly. They must be intertwined at all times. (p. 342)

During the same congress, Hans Freudenthal, who was its president, suggested that I organize a round table devoted to the psychological problems of mathematical education. Lee Shulman joined me in fulfilling the task, and the round table took place. Although the meeting had not been prepared in advance, it represented a real success. Consequently, it was decided that at the Second International Congress on Mathematics Education (in Exeter, Great Britain, in 1972) a workshop would be held that was devoted to the psychology of mathematics education. I had the honor of chairing that workshop. This time, speakers were invited to present papers. The workshop was a great success. Hundreds of participants attended. As the proceedings of the congress noted, "this was the most popular group" (Howson, 1973, p. 15). It is possible that many attended the sessions initially out of sheer curiosity, but very soon all the participants became aware that the psychological problems of mathematical learning and reasoning are scientifically exciting and at the same time genuinely relevant for mathematics education.

At the Third Congress on Mathematical Education (in Karlsruhe, West Germany, in 1976), as a conclusion of the second workshop devoted to the psychological problems, it was decided to create a permanent body, the International Group for the Psychology of Mathematics Education (IGPME, which later became known as the PME group). It was also decided that the members of the group would meet every year. I am pretty well convinced now that our organization has represented a real success from various points of view: numerically, scientifically, and socially. At the beginning, a few countries were represented at the PME conferences (West Germany, Great Britain, France, The Netherlands, Canada, Israel). At the most recent conference (in Hungary in 1988) most of the European countries were represented. In addition, we had North and South Americans and colleagues from Asia, Australia, and Africa. A large variety of problems were discussed, from intellectual skills to affective influences and aspects of computer environments. The existence of the PME has stimulated new research perspectives, international collaboration, and the development of new theoretical frameworks.

There is one basic explanation for this success: Mathematics educators realized, beyond any doubt, that mathematical activity cannot be restricted to purely formal constraints that act in a vacuum (as many university professors, authors of new curricula, seemed to have assumed). People involved in mathematics education began to realize that, like every intellectual activity, mathematical activity is dependent on an enormous variety of constraints--motivational, affective, imaginative, linguistic, and so on. We started to discuss conflicts and inconsistencies, relational and instrumental understanding (Skemp, 1979, pp. 259-264), intuitions and models, structures and functions, concept images and concept definitions, conceptual fields, epistemological obstacles, and representations and symbols. More and more, psychological concepts and psychological research methods became involved in innovative attempts in mathematics education. This effervescence of ideas was contagious. Mathematics professors who did not believe in educational research became at least receptive to its suggestions. Schoolteachers would like to get more practical, useful information from research journals. Briefly speaking, there is an increased need all over the world to learn more about the psychological texture of mathematical activity. Certainly, the PME group has made an important contribution to that development. Its creation has been a success because it responded to profound theoretical and practical needs. I believe that today we may be proud of our achievements.

As a result of the creation of PME, a fundamental change, more remarkable now in retrospect, has taken place in attitudes about investigation. The mathematics education journals used to publish two types of papers according to the respective profiles of the journals. Some of the papers expressed the authors' suggestions concerning the teaching of mathematics: new topics, new examples, new ways of teaching. No empirical data were, generally, invoked for supporting the authors' ideas. A second category of papers published in research journals presented, usually, applications of psychological problems, concepts, and theories to the mathematical field (cognitive styles, sex differences, problem solving, the role of imagery, etc.).

The creation of PME changed the perspective and the approach to investigation. More and more, psychological problems *inspired by the school reality* captured the interest of researchers: computation, numbers, fractions, proportional reasoning, elementary arithmetical operations, visualization, geometry, the use of symbols, mathematical proofs, computer environments, and so forth. But higher order mathematical concepts have penetrated only slowly into the field of investigation. Concepts like infinity, limit, functions, and graphs appear rather seldom. Very few papers refer to probability and combinatorics. In fact, most of the more complex mathematical concepts are still insufficiently investigated from the psychological point of view.

New Trends in Mathematics Education

Let us consider briefly some of the new trends that tend to shape the future image of mathematics education:

1. Especially during the last 15 years researchers have started to devote much time and effort to the problems and concepts of so-called artificial intelligence. They study the role of computer environments; they try to translate the problems of mathematics education in terms of the information-processing approach. Developmental and cognitive psychology have been replaced as major sources of influence by the more

modern perspective of *cognitive science*. We talk about short-term and long-term memory, about working memory, and about routines and subroutines.

From various points of view, artificial intelligence metaphors and, particularly, the use of computers in instruction, may have beneficial effects, both theoretically and didactically. By learning to have a dialogue with the computer, students may improve their capacity to analyze their own reasoning processes. Students may become involved in an active, intelligent interaction with the computer in defining and solving original problems.

But certainly the use of the computer in the instructional process has its limitations. If these are not understood, the computer becomes a real danger. Unfortunately, and I would like to emphasize this point especially, we are witnessing nowadays the penetration of computers at every level of instruction *without a serious research basis and without systematic attempts to evaluate their psychological and didactical effects.*

In the opinion of H. L. Dreyfus and S. E. Dreyfus (1986), expertise cannot be reduced to an aggregate of elementary skills acquired stepwise at inferior levels. In fact, these authors claim, the expert reacts globally, by intuition, on the basis of rich experience, and formerly learned rules do not play an important role any more. By insisting, at the expert level, on sets of analytically accountable elementary skills (as has been done with the beginner), one only hampers the development of the expert's solution capacity. In other words, in the view of Dreyfus and Dreyfus, computer-assisted instruction may be useful for nonexpert performance, but may play a negative role at the level of expertise.

I am insisting on that point because the present tendency is to include the computer in teaching programs *without a careful consideration of the ensemble of its psychological and didactical effects.* Personally, I am not inclined to agree that intuitively based reactions are not rational, as Dreyfus and Dreyfus do. On the contrary, my assumption is that intuition is also scientifically analyzable (see Fischbein, 1987). But this does not imply that we are able to confer on the computer the task of simulating every type and every level of our intelligent behavior and that the computer may tutor every aspect of a learning process.

I think that it should be a major task of PME to initiate research projects aimed at producing experimental and observational data related to that area. For instance, an instructional computer program, however "intelligent" it may be, necessarily restricts the learner's horizon to a number of fixed constraints. Does not this situation impair the students' creative perspective? Another major problem is represented by the effects computer-assisted instruction may have at various age levels and on various types and levels of intelligence.

These and many other problems should be considered if one wants to prevent the computer from causing more harm than good as an instructional tool.

2. The constructivist approach is another source of fruitful ideas. As has been pointed out, constructivism is not an ontology, a theory of existence, but a theory of

knowledge. Our cognitions are not duplicates of an assumed external world but rather constructions whose aim is to guarantee the practical success of our behavior. In the words of von Glaserfeld (1986) "from this changed point of view, then, the cognitive activity does not strive to attain a veridical picture of an 'objective' world but it strives to obtain viable solutions to whatever problems it happens to deal with" (p. 109).

Neither philosophically nor psychologically is this a totally new conception. It is reminiscent of the pragmatist philosophies and of the constructivist ides of Jean Piaget. But mathematics education only recently became aware of its didactical implications. A possible implication is that one has much to learn from the child's own behavior when he or she tries to cope with mathematical concepts and operations. How do children count? How do they add, subtract, multiply, or divide? The traditionally accepted hypothesis was that the teacher teaches the child and that children understand and assimilate more or less correctly what they have been taught. The new point of view is that children invent their own methods of counting, adding, and so forth; that these methods may be totally different from those usually presented by the teacher; and that children's methods may be more adequate to their own way of thinking than those proposed by the teacher. Consequently, one should devote much more time to observing the child's spontaneous behavior when coping with mathematical problems and to try to adapt the teaching methods to those spontaneously used by the child. (For a critical analysis of constructivism, see the paper by Kilpatrick, 1987).

There are some aspects related to the constructivist approach that should be discussed and that require serious investigation. First, does the child's mind progress naturally, spontaneously towards the formal ways of reasoning required by a mathematical activity? Piaget's stage theory seems to support such a claim. But are things really so? There is much evidence that every important progression in the child's reasoning capacity, every acquisition of new statements, may be achieved only as an effect of practice. The adolescent would never acquire spontaneously the formal perspective imposed by mathematics, the capacity to rely on pure formal constraints, or the need for an axiomatic structure of a mathematical system of statements and proofs. On the contrary, our natural tendency is to rely on behavioral hints, pictorial models, and intuitive justifications. To learn mathematics means to *construct* mathematics. Mathematical activity is essentially a constructive process. The student learns mathematics not by absorbing concepts, definitions, theorems, and proofs but by constructing them through his or her own intellectual efforts. But individuals usually do not do all these things by responding to their own problems and by resorting to their own natural, intellectual means. Our natural behavior is adapted to the concrete reality in which we live and not to formal constructs governed by formal definitions and rules.

The task of the teacher is to create an environment that would require a mathematical attitude, mathematical concepts, and mathematical solutions. Let me take a single example: How many groups of three letters may one produce using the letters A and B (with replication)? Ten-year-old children do not have a spontaneous answer to this question, but they may start to construct it by combining the various letters according to the requirements of the problem. They are now halfway toward the answer. Building the groups, they exercise their constructional ability, which is part of their mathematical capacity. The next step would be to produce a *method* that would allow them to find all the groups systematically. The problem of finding a method, an algorithm used consciously represents a fundamental aspect of mathematical reasoning. Should the teacher wait until children find the method by themselves without any help? This is a crucial question for mathematics education. As I have already said, in my opinion formal reasoning does not develop spontaneously as a main way of thinking. This conclusion does not imply that the teacher should simply offer the solution. What the teacher should do is to *direct* the student's efforts to a solution by asking adequate questions. The student *builds* the answers as a reaction to a certain environment. This environment should be programmed as a problematic one in order to inspire the student's solution endeavors. For instance, the teacher can prepare the way for the building of a tree diagram in response to a problem, continuing to ask at the initial steps, until the construction principle is grasped by the child: "What are the possibilities now?" The formula for the above example will then, finally, be obtained as an effect of a constructive process. What the constructivist theory has not emphasized so far is that the student acquires this way more than a formally consistent procedure or mathematical relation. What he or she gets by the constructive process is a personal involvement leading to a genuine insight, a genuine intuitive understanding of the respective concepts and procedures.

Briefly speaking, in my opinion a correct constructivist approach implies the creation of a problematic environment that would elicit the student's adequate constructive endeavors. An "intelligent" computer program, well integrated into the curriculum, might play an important role in that direction. But this use would represent only the initial level of a constructive instruction process. A higher level would be attained when students could grasp and formulate by themselves certain problems inspired by the given conditions.

3. A third theoretical direction that deserves more attention refers to the relationships between the formal, the algorithmic, and the intuitive aspects of mathematical activity. At every level of a mathematical activity these three aspects are present. The formal level is represented by the definitions, theorems, and proofs related to certain concepts and properties. The algorithmic level is represented by the various solution procedures. The intuitive component is expressed in the intrinsic beliefs attached to concepts and operations. In certain cases, these aspects may act in harmony, but very often there are discordances and conflicts. A student may be able to utter correctly a definition or a theorem, but his tacit belief, his tacit interpretation, may be dissonant with respect to the mathematical truth. This creates

a complex dynamics that should be elucidated as a preliminary condition for genuine understanding.

4. These observations bring us to the concept of metacognition. The term means briefly a person's cognition about cognition.

> The premise behind research in this domain is that persons are not only organisms who cognize about objects, events, and behaviors; but importantly they also cognize about cognition itself. They form and hold conceptions about how the mind works, about which mental problems are hard, which easy, about their own mental states and processes. (Wellman, 1985, p. 1)

In my opinion, one has to go a step further. *One has to develop systematically this reflective capacity* and to endow students, starting from adolescence, with the intellectual means for monitoring their own reasoning processes. An essential point would be to help students to identify the sources of their misconceptions and their systematic procedural errors. Rigid associations, inadequate models or an inadequate handling of models, inadequate intuitive beliefs, incorrect generalizations, and so forth should be identified and the student made aware of their impact. This is an enormously complex task. It implies, first, a theory of mathematical errors. The present data related to *bugs* (analogous to computer bugs) are a good contribution but totally insufficient. Second, it implies research projects aimed not only at identifying systematic errors but also at explaining them, determining their deep causes. And finally, it implies research projects aimed at checking and developing the student's capacity to understand the real causes of his or her own errors and to control their impact. Classroom and small group interactions may play an important role in making students aware of the contradictions to which they led by their inadequate models and beliefs (see Schoenfeld, 1987).

The Need for a Theoretical Background

Generally speaking, I feel that the time is ripe now for a collective endeavor to develop a theoretical background for our field.

Initially, the tendency was to borrow questions, concepts, theories, and methodology from psychology. Associationism, behaviorism and neo-behaviorism, gestaltism, the Piagetian school--all of these views have had a certain tacit or explicit impact on mathematics education (theory and research). But the real practical profit was rather slim. The explanation is that psychology is not a deductive discipline. The mere application of general principles to particular domains does not usually lead to significant findings. Even cognitive domains strongly related to mathematics education--like problem solving, memory, reasoning strategies, creativity, representation, and imagery--cannot result directly in practical, useful suggestions for mathematics education and cannot represent by themselves the major sources of research problems in this field. Even Piagetian stage theory and the respective findings concerning mathematical concepts (number, space, chance, function, etc.) cannot be translated directly in curriculum terms.

This observation does not mean that mathematics education should live and develop in a closed shell, opaque to external influences. Psychological and sociological coordinates are valuable prerequisites for defining problems, devising research projects, and interpreting the data. Nevertheless, these prerequisites are by themselves totally insufficient.

Mathematics education raises its own psychological problems, which a professional psychologist would never encounter in his own area. I do not see a psychologist becoming interested in the specific types of representations that appear in mathematics--from graphs representing functions, to various classes of morphisms, to the dynamics of mathematical symbolism (see Janvier, 1987). I do not see how a cognitive psychologist would become interested in and would handle the problems raised by mathematical infinity with all its various facets and difficulties. And so on. In order to be able to cope with such problems, one needs a particular system of concepts in addition to the general ones inspired by psychology. Even usual psychological concepts acquire new meanings in the light of mathematics and mathematics education. Let me mention some concepts invented as an effect of that endeavor: concept image and concept definition (Vinner and Tall), conceptual field (Vergnaud), and primary and secondary intuitions (Fischbein). Traditional psychological concepts like problem solving, representation, understanding, and learning get a new, much richer content in relation to mathematical activity.

The relationships between abstract and concrete, between formal and intuitive, between algorithmic and heuristic present such a variety of novel facets in the light of mathematics that no deductive endeavor can predict them from general psychological concepts. This inability means that, in fact, cognitive psychology itself may be enriched with new meanings, categories, and paradigms inspired by ideas and findings produced by mathematics education research.

Further, the research methodology should also be adapted to the specificity of the domain. We are used, in psychology, to highly complex and carefully devised statistical procedures. One tends to define with maximum accuracy the variables involved; one tends to obtain quantitative results whose degrees of confidence may be measured precisely. All these procedures imply the assumption that the object under investigation is really analyzable into elementary components (as is the case with physics or chemistry), components that can be handled mathematically and that, when regrouped, will reproduce the original structure--or another structure, according to our aims (in our case, didactical aims). Consequently, the psychologist is inclined to choose problems and situations that would allow such an analysis without violating the genuine nature of the phenomenon. Elementary perceptions; the learning of very simple items; solving very simple, clearly definable problems; and elementary reactions to determined stimuli are examples of such topics. But mathematical activity cannot be reduced to such extremely simplified aspects. Psychological research related to mathematics education must combine a large variety of research methods including classroom observations, dialogues with teachers, clinical methods, interviews, questionnaires, case studies, experimental lessons, and historical analyses. Pure statistics based on laboratory-like investigations will always yield only incomplete, superficial information. As a matter of fact, it is this comprehensive direction that is more and more represented in our domain, especially during the last decade. The book edited by Richard Shumway (1980) is a useful instrument, but we still lack a sufficiently comprehensive, up-to-date monograph referring to research in mathematics education in which the contribution of the psychological perspective was systematically taken into account. This lack imposes the necessity of working out a comprehensive, systematic theory of research methodology in mathematics education. Contributions in that direction are almost absent. As a result, one may still encounter studies that are excellent from a formal point of view but that lack any real theoretical or didactical significance, and there are many studies containing interesting ideas that are totally unsatisfactory from the methodological point of view.

The Quest for a Paradigm

As a matter of fact, a theory of research methodology can be produced only in the context of a general theoretical perspective of our domain. Such a theory is still absent. Can any one of the present trends--the constructivist approach, the computer metaphor, the information-processing conception, the intuitive-formal dialectic, the sociological and cultural interpretation--play the role of a real paradigm for mathematics education? (For more details about the present trends and objectives in mathematics education, see Romberg and Stewart, 1987).

Let me take the meaning of the term *paradigm* as it is described by Thomas Kuhn (1970). Kuhn mentions some examples of remarkable scientific achievements that played the role of paradigms in the history of science (Newton's Principia and Optiks, Franklin's Electricity, etc.), and he writes:

> Their achievement was sufficiently unprecedented to attract an eduring group of adherents away from competing modes of scientific activity. Simultaneously, it was sufficiently open-ended to leave all sorts of problems for the redefined group of practioners to resolve.
>
> Achievements that share these two characteristics I shall henceforth refer to as "paradigms" a term that relates closely to "normal science". By choosing it, I mean to suggest that some accepted examples of actual scientific practice--examples which include law, theory, application and instrumentation together--provide models from which spring particular, coherent traditions of scientific research. (p. 10)

In my opinion, none of the present trends in mathematics education (including the investigation of psychological aspects) constitutes a genuine paradigm. The constructivist approach is a theoretical perspective, but it still lacks "law" and "instrumentation." The applicative schemata are not yet clarified. The theory that emphasizes the dynamics of the formal-algorithmic-intuitive aspects of mathematical reasoning has had so far only a tiny impact on mathematics education research. The information-processing approach, though very stimulating for cognitive research, does not seem to encompass the entire variety of the psychological aspects of mathematics education. It still needs a lot of theoretical elaboration. The related computer metaphor attracts the interests of many researchers involved in mathematics education, but it lacks a solid theoretical basis and valuable feedback from school applications. One may mention, in addition some other lines of investigation--like those related to the concept of schema, those insisting on the language model (Nesher), the novice-expert comparisons, or the story-shell curriculum units (Romberg). All these constitute promising attempts in which psychological arguments are present. But none of them has proved itself a sufficiently powerful synthesis between theoretical perspectives and practical tools as to become a dominant, largely accepted paradigm. (For the philosophical debate concerning the theory of mathematical education, see Orton, 1988).

Concluding Remarks

The creation of PME in 1976 was the natural effect of the need felt by large circles involved in mathematics education for more systematic, specifically adapted psychological information. Nowadays, hundreds of researchers all over the world raise problems, investigate, theorize, publish, and confront their findings in the relatively defined area of the psychology of mathematics education. That domain does not constitute any more a conglomerate of unrelated attempts to apply psychology to the teaching of mathematics. Many problems and some new categories and theoretical views have been suggested by mathematics education itself. We do not yet possess a particular dominant paradigm, but one may nevertheless affirm that the *psychology of mathematics education* tends to become the paradigm of mathematics education (as a scientific body of knowledge) *in general*. My global evaluation is that most of the research going on nowadays is related to the psychological aspects and the activity of the PME group.

There are some basic directions to which attention ought to be focused in the future. First, we ought to enlarge the realm of "home grown" problems, particularly with questions related to concepts of advanced mathematics such as classes of numbers (especially irrational and complex numbers); infinity (dynamic and actual infinity, the concept of limit); the concept of function; the basic ideas of calculus (the concepts of derivative and integral), infinite sequences and series, probability and statistics; geometry (Euclidean and projective geometry, topology, analytic representations); and mathematical proofs and the axiomatic method. These and many other concepts associated with intuitive obstacles are only insufficiently represented in our research projects. Second, there is an absolute need for symposia and working groups debating theoretical problems. Let me mention again the information-processing approach (Is this theoretical model able to cover the entire domain of the cognitive processes involved in mathematical reasoning?); the need for a more specific definition of constructivism as a psychological model for mathematical education (For instance, what should the role of the teacher be?); the limitations of the computer metaphor;

the dynamics of the formal-algorithmic-intuitive aspects; the role of social interactions in developing mathematical reasoning. There is an urgent need for a systematic debate concerning research methodology, that is, the field of various investigative techniques, their limitations, and their mutual complementarity.

Such theoretical analyses (both conceptual and procedural) should take into account the dual, dialectically contradictory nature of mathematical concepts, operations, and proofs. Namely, their strict formal character when considered as end products and their profound association with empirical, motivational, affective, and intuitive aspects when considered in the light of mental endeavors.

1 EPISTEMOLOGY AND PSYCHOLOGY OF MATHEMATICS EDUCATION

Gérard Vergnaud
with the collaboration of
George Booker, Jere Confrey, Stephen Lerman, Jack Lochhead,
Anna Sfard, Anna Sierpinska, and David Wheeler[*]

Main Questions

Epistemology is concerned with one main question: What is knowledge?

From this question, many other questions can be derived: How is knowledge acquired? What are the parts played by action, by perception, by language and symbolism in the development and the functioning of knowledge? What is the relationship between routinized knowledge and problem solving? And so forth.

There are also epistemological questions that are specific to mathematics: What kind of objects is mathematics all about? Are there different kinds? What is the relationship of mathematics to other sciences and to other fields of human experience? In what sense is mathematics both a set of tools and a set of objects?

To many researchers such questions appear philosophical, nonempirical, and possibly useless. But it is easy to trace implicit epistemologies in researchers' work and in the way teachers teach. Therefore it is wise to try to clarify, as much as possible, the main epistemological issues that can be raised and state explicitly our own standpoints. For instance, some teachers think that mathematics is timeless truths, waiting to be discovered like an unknown country; others think that mathematics can be completely reinvented by students. An important epistemological debate concerns the part of ordinary experience and the part of physics and other disciplines

---

[*]Acknowledgments go also to Alan Bell, Willibald Dörfler, Tommy Dreyfus, Ted Eisenberg, Carolyn Kieran, Colette Laborde, and Pearla Nesher, who criticized an earlier version of this chapter.

in the meaning of mathematical concepts; another concerns the part of symbolism and formalism.

There are at least three levels of questions interfering: epistemology of mathematics, epistemology of psychology, epistemology of mathematics education.

The epistemology of mathematics consists traditionally of several approaches: One stems from the spontaneous reflection by mathematicians themselves on the nature of their knowledge and on the nature of the invention and discovery process. Several great mathematicians have endeavored to clarify the status of mathematical thinking, like Poincaré (1913, 1920) and Hadamard (1949) in France. Another approach is historical; its aim is to understand the scientific and social environment in which new mathematical concepts and techniques have emerged and developed. This approach can be found in the book by Davis and Hersh (1981) and in many other books. Finally, the mathematico-philosophical debate on the foundations of mathematics that has taken place during the first half of this century about logicism, intuitionism, formalism, constructivism, and so on (Benacerraf & Putnam, 1964) has invaded the whole field of epistemology of mathematics.

The epistemology of psychology has different roots. One debate concerns the nature of objects that psychology is concerned with, as a science: behavior? consciousness? unconsciousness? elementary modules of action, perception, and memory? complex organizations of behavior? complex representations? Another debate concerns the kind of models that can be used to give an account of psychological phenomena: associations? inborn mechanisms? general biological processes like adaptation? neurophysiological models? linguistics? computer science models?

The epistemology of mathematics education inherits questions from both fields (mathematics and psychology) and adds new ones because mathematics education takes place in a certain society, a certain institution, a certain classroom, with such different aims as the education of future mathematicians and the education of rank-and-file citizens. These social constraints on mathematics education do not modify the nature of mathematical knowledge per se, but they have strong implications for the way teachers see the teaching of mathematics and mathematics itself. Not only do students' representations of mathematics differ from those of teachers, but also teachers' representations vary a lot, according to their views of mathematics, psychology, and society.

It is impossible to deal with all these questions in a short chapter. Therefore we devote our attention to a restricted set of epistemological questions that are central both to the study of the learning-rediscovery-reinvention process in students' minds and to the history of mathematics: What is the nature and the function of a new concept, a new procedure, a new type of reasoning, a new representation? More precisely, what is the relationship of new mathematical competencies and conceptions to the practical or theoretical problems that make them useful and meaningful?

This kind of question is essential for the choice of situations by teachers, inasmuch as the identification of the concepts involved and the relevant properties of theorems is crucial for the cognitive analysis of tasks and behaviors--especially for the analysis of novelty.

It is also the kind of epistemological question that drives the historian's inquiry when he or she tries to find out the scientific and social circumstances under which mathematical inventions have emerged. There is much to be gained from an interactive study of individual and historical processes of developing mathematical knowledge. The study of obstacles met by mathematicians in the past helps us to interpret the errors made by students today; in turn, the study of students' errors, difficulties, and wrong conceptualizations sheds some light on our understanding of the history of mathematics (Arsac, 1987; Sierpinska, 1985). Even if the set of problems that students may profitably meet is different from the set of problems that scientists have met in the course of history, it is essential for the psychology of mathematics education to consider the relationship of knowledge to problems.

In many psychological studies on mathematics education, researchers take it for granted that students should learn such a topic, at such a level, with such a method, without questioning the framework within which these choices take place. One does not usually question the fact that addition is a binary operation, that one does not teach directed numbers at the elementary level, or that one introduces algebra using the natural numbers only. Consequently one misses (and students miss) some very important points concerning the meaning of addition, directed numbers, and algebra.

Epistemology is not concerned only with the foundations of mathematics, as the debate on the possibility of reducing arithmetic to logic (Carnap, 1931/1964; Gödel, 1947/1964; Russell, 1919) or on intuitionism (Brouwer, 1913/1964; Heyting, 1956) might suggest. Epistemology is needed on all floors of the mathematical building every time we need to clarify the relationship of mathematical knowledge to problems. For instance, the concept of volume is made of different properties and different relationships to other mathematical concepts: There are elementary properties of the concept that can be grasped by 6- or 7-year-olds (in such tasks as the comparison of containers, the addition and subtraction of volumes, and measurement with a unit of volume), whereas some three-dimensional properties of the volume of right parallel-epipeds and prisms (like the relationship of ratios between lengths, areas, and volumes) are not yet mastered by the majority of 15- to 16-year-olds. Different problems involve different properties of the same concept and different concepts.

The relationship of knowledge to the problems to be solved and the situations to be dealt with makes it relevant to consider also the influence of the cultural environment on the learning and teaching of mathematics. Mathematics education is a social process that takes place in different cultures and different societies with different school organizations, different philosophical backgrounds, and different goals. The meaning of mathematics education is different in a highly developed technological society from that in a traditional rural society. It is also different for different subgroups of the same society. Mathematics is becoming important for everybody: We need some mathematics to understand computer-assisted machinery, accountancy, and conventional technology, and also to make sense of the information that is delivered to us by the mass media. These requirements raise the problem of the aims of mathematics education. If mathematics is not only a part of our scientific culture but also a useful set of competencies and conceptions for a variety of professional and other activities, then we must revise the overall philosophy of mathematics education, especially at the primary and secondary levels.

We need to question the implicit mental representations that teachers, parents, employers, decision makers, and researchers have about mathematics and mathematics education. We need to question curricula and not take them for granted. We need to question schoolbooks; software materials; and teachers' habits, conceptions, and decisions, from the point of view not only of their sociological and psychological adequacy but also of their epistemological foundation. For instance, the use of computer programs makes it necessary for many people to understand the concept of function at a level of generality far above an understanding of the covariation of goods and costs, or mass and volume.

We next report and comment on a few important epistemological points of view concerning the psychology of mathematics education. They range across a wide theoretical spectrum, from Piagetian theory to the artificial intelligence approach. Then we outline a possible synthetic framework and try to illustrate a few important theoretical ideas by taking the example of the long-term development of the concept of number. Finally, we draw some perspectives for mathematics education and future research.

Some Contributions to Epistemology from a Psychological Perspective

Our starting point will be Piaget, as he is probably the first and most important contributor to the epistemology of mathematics from the point of view of psychology. He made many contributions to the ontogenesis of mathematical and logical structures with the help of collaborators such as Inhelder, Szeminska, Greco, and Vinh Bang. This chapter is not the place to report his work on the concepts of quantity and number, representation of space, concepts of chance and probability, or logical operations concerning classes and propositions. But it is important to note that Piaget used the word *epistemology* repeatedly; for instance, in his three-volume book *Introduction à l'Épistémologie Génétique* (1949) and in the collection *Etudes d'Épistémologie Génétique*. It is difficult to summarize his views in a few paragraphs, but one may consider the following ideas as essential.

1. Knowledge derives from the adaptation of the individual to his or her environment. This thesis reflects the background in biology of Piaget, who considered the process of knowing to be a particular case of the assimilation and accommodation process that characterizes all living organisms: assimila-

tion of new situations and new objects to former structures, and accommodation (modification) of these structures to the new characteristics of objects.

2. Knowledge can be traced to the individual's way of acting with objects and dealing with situations and not only to his or her declarations. Action is the main factor in the knowing process. Language was seen by Piaget as both a consequence and a factor of thinking, but Piaget stressed action more than language. Vygotsky placed the stress differently.

3. When acting on objects, individuals develop different kinds of knowledge, depending on the kind of abstraction they make: Empirical abstraction consists in isolating the properties and relationships of external objects, whereas reflective abstraction consists in isolating the properties and relationships of the person's operations themselves.

4. For Piaget, logico-mathematical knowledge derived from reflective abstraction, whereas physical or biological knowledge came from empirical abstraction. We discuss this point below.

5. Although Piaget studied the development of specific mathematical concepts extensively, he attempted to give an account of the development of intelligence and knowledge in purely logical terms: For instance, he tried to characterize the "concrete stage" and the "formal stage" of cognitive development as sets of logical operations. Nevertheless, his findings were not content free, and he was faced with the *décalage* problem: that is, when the same logical structure does not apply equally well to different objects or to different aspects of the same objects. For instance, there is a décalage between the conservation of the substance (quantity of matter), the weight, and the volume of the same piece of plasticine.

6. Symbolic activity is the inner counterpart of overt activity and results mainly from the interiorization process by which overt activity (motoric, perceptual, communicative, etc.) is progressively erased and made internal (Piaget, 1945). For instance, the overt activity of counting a set of objects can be progressively interiorized: Gestures become less and less visible, number words less and less audible. This point is important in Piagetian theory, as it leads to the thesis that thinking is not merely interiorized perception or interiorized language but also, and rather essentially, interiorized action. A concept is operational, or it is not.

It is interesting to evaluate Piaget's ideas in the light of present research. His "constructivist" view of mathematical competencies and conceptions (knowledge is produced by the personal activity of the child) is probably one of the most widely accepted today among researchers in the psychology of mathematics education, even though many of them complain that Piaget did not pay enough attention to the social aspects of the teaching-learning process and did not integrate the important views that have been developed by Vygotsky, Bruner, and others. Actually, Piaget never studied the teaching-learning process, either in the classroom or in the home. It is one of his omissions, and an important burden for present and future research on mathematics education.

Another unclear point in Piaget's theory is the relationship of the knowing process to the physical and social characteristics of the situations that students are faced with. Teachers play an important part not only by explaining, showing, and managing the classroom situation but also by choosing carefully and adequately the situations that make mathematical knowledge meaningful. The relationship between problems to be solved and specific competencies and conceptions is not really clarified in Piaget's work. This is probably the point to which present research has made the most striking contribution: to the early concept of number, additive and multiplicative structures, algebra, geometry, programming, and the concepts of limit, convergence, and continuity (see the subsequent chapters of this book).

Last but not least, one may find to be misleading Piaget's thesis that mathematical knowledge comes from abstracting the properties and relationships of operations and not of objects. This view prevented Piaget from understanding why the early concept of number is inextricably tied to the concept of measure (cardinality of collections, measure of spatial magnitudes, measure of mass, etc.). It is certainly true that numbers gain autonomy in the course of cognitive development and get rid, to some extent, of their physical meaning. It is true also that geometry has to do with purely spatial objects, properties, relationships, and transformations. But neither number nor geometry would exist if they were not acceptable models of physical entities and did not help in dealing with empirical problems. Mathematics is a way to conceptualize the real world, at least during the early stages of mathematics education.

Probably no other psychologist deserves as much attention as Piaget concerning the topic of this chapter. But interesting enough are some alternative or complementary views of the relationship between the epistemology and the psychology of mathematics education.

Vygotsky (1962), being interested in the role of symbols and language in the development of thinking, focused some attention on the relationship between the implicit knowledge present in children's arithmetic reasoning and the explicit knowledge required to understand and use algebra. He even drew a parallel between the move from arithmetic to algebra and the move from spoken language to written language, as well as from the mother tongue to foreign languages. In all three cases, Vygotsky said, one can see the need for a transformation of the implicit knowledge of rules and structures into the explicit formalization of these rules and structures. This way of looking at arithmetic and algebra does not give an account of all the interesting differences between arithmetic and algebra, but it points to the fact that symbols function in algebra very differently from the way words and digits function in arithmetic. Abstraction is needed to identify as objects not only the magnitudes involved but also their relationships, and to represent them by formal expressions.

Another important difference between arithmetic and algebra, which Vygotsky did not mention, is that algebra offers students the possibility of solving problems that they would otherwise fail to solve--problems involving several unknowns, for instance. Algebra also requires students to identify new mathematical objects like those of variable and function, monomial and polynomial, and structured sets of numbers (see Kieran, this volume).

The specific function of symbols in algebra raises interesting problems from an epistemological point of view, as it exemplifies the power of modelling: to extract functional relationships as a model of the situations, to operate within the model without paying attention to the external meaning of such operations, and to interpret afterwards the result of such operations. The same function appears in other symbolic systems like tables, diagrams, and graphs. But this specific function of symbols does not give a full account of the part played by language and symbols in thinking, namely conceptualization.

As a matter of fact, a great part of our knowledge, as can be traced to our behavior, is only implicit: We take up information with the help of invariants (categories, relationships, and higher level entities), without expressing or even being able to express these invariants. This is especially visible in students' mathematical behavior, as they often choose the right thing to do without being able to mention the reasons for it. The cognitive analysis of such behaviors very often reveals the existence of powerful implicit mathematical concepts and theorems, as we show below. Let us call them concepts-in-action and theorems-in-action. Such knowledge cannot be properly called "conceptual," as conceptual knowledge is necessarily explicit. Therefore words and symbols, sentences and symbolic expressions are indispensable cognitive instruments for the transformation of implicit operational invariants into concepts and theorems. The most important cognitive function of language is to contribute to the identification of relevant features as objects. This is probably what Vygotsky had in mind, even if he did not express it that way.

It may be interesting to contrast these views with alternative epistemologies of mathematics education; for instance, with the formalistic conception, the intuitionistic point of view, or the artificial intelligence approach. Contrast is not contradiction: There are probably complementary aspects in those different perspectives.

It is important to note, first, that the formalistic and the intuitionistic epistemologies concern mathematics itself and not psychology, whereas artificial intelligence has had little influence on mathematics up to now and a great influence on psychology. We will not discuss the formalistic conception in mathematics here but only its derivative in mathematics education. Its influence was very strong 20 years ago, probably more in some countries (like France) than others. It is certainly less influential now.

The formalistic conception requires that everything be expressed formally, without ambiguity; the ultimate goal of mathematics is to reduce mathematical truth to the syntactic coherence of formal and symbolic systems. If mathematical ideas may be implicit, then the formalist conception goes astray when it requires that everything be expressed formally. If mathematical ideas are not concepts until they are expressed, then formalists are right in trying to get from students, at least from advanced students, explicit definitions and rules, well-formed formulas, and well-defined conditions of validity. What is wrong in the formalistic view is its blindness to the fact that mathematical ideas grow and change over a long period of cognitive development, through a variety of situations and activities, and that formal and axiomatized knowledge can only be the last-developed state of a student's knowledge, the small visible tip of the iceberg. The issue is whether mathematics can be taught with definitions and axioms or must be presented as a way of conceptualizing a wide range of situations through problem solving, and with the help of progressive and partial "explicitation." The meaning of mathematics comes essentially from problems to be solved, not from definitions and formulas. Yet it would be wrong to deny the importance of strict definitions and formulations in mathematics when the problem to be solved is to state theorems and express conditions unambiguously, in order to get other mathematician's agreement or to make an automatic (and therefore syntactic) model of the field.

The intuitionistic point of view in the psychology of mathematics education, as best expressed by Fischbein (1987), has no direct connection with the work of mathematicians like Brouwer and Heyting. But Fischbein argues similarly, in contrast with the formalistic view, that thinking would simply be impossible if we could not rely on immediate, self-evident intuitions. Fischbein describes different characteristics of intuitions: It is difficult to act against them; they last for many years; they are general and global; they are necessary for action. He proposes a classification of intuitions and analyzes some mechanisms like overconfidence, premature closure (a halt to the search for new information), and the primacy effect (giving privilege to first interpretations). Fischbein provides examples of correct mathematical intuitions (transitivity of equivalence relationships) and overgeneralized intuitions (multiplication makes larger and division smaller). He also gives examples of primary intuitions (the more space, the more things) and secondary, or constructed, intuitions (conservation of quantity). In a sense, Fischbein travels the same road as Piaget and Vygotsky. Yet a theoretical problem is that psychologists cannot take intuitions for granted but rather must try to analyze their content and their development. Intuitions change in the course of cognitive development: The intuition of subtraction is different for the 5-year-old and the 16-year-old, as is the intuition of similarity. The feeling of immediacy that we have with our intuitions is therefore misleading and needs to be analyzed as the result of accommodation to experience and as a change in implicit knowledge. It is difficult to think of a mathematical intuition that would not rely on some experience and therefore some discovery or construction, however conscious or unconscious that construction might be.

It is precisely the need to make a model of implicit thinking that has led to the development of the information-processing approach in psychology. This approach has been fairly successful in some parts of psychology like the study of perception but much less so in the study of complex behavior. The field of application of artificial intelligence models that has proved to be the most fruitful for the study of mathematical behavior is probably the field of calculations (addition, subtraction, multiplication, division), and more generally the fields in which one can identify algorithms or quasi-algorithmic procedures and therefore try to guess the rules that are used by students in place of the correct ones. Every time behavior consists of a sequence of choices among a limited range of unambiguously identified possibilities, the "production-rule" paradigm can work fairly easily. When the choice relies on the conceptualization of a new object, or a new relationship, the production-rule paradigm fails.

This is a basic epistemological weakness of information-processing models: They do not provide any theory of what a concept is, and especially of its operational character (i.e., its relationship to a variety of problems). For instance there is no convincing model of the development of the concept of number and of the theorems associated with additive and multiplicative structures.

The artificial intelligence approach also does not offer any plausible theory of the part that language and symbols play in thinking. Either it identifies and reduces thinking to symbol manipulation, or it deals with implicit thinking as if it were a set of unambiguously named objects.

Last but not least, it does not offer any plausible view of the long-term development of students' competencies and conceptions that takes place through interactions with problems and with other individuals, therefore through action and communication.

Epistemological considerations on the nature of concepts are probably the best way to detect the weakness of a theory or a model. Epistemology cannot be married with overly simplistic theories of behavior and thinking.

Most psychologists interested in mathematics education research today are in some sense constructivists. They believe that competencies and conceptions are constructed by students themselves. But the fact that most researchers do not specify enough the physical and social conditions under which knowledge construction takes place opens the way to a wide range of epistemological positions, from radical constructivism, which denies the possibility of the mind to reflect objective aspects of reality and minimizes the part played by teachers (Cobb, 1986; von Glasersfeld, 1983), to social constructivism, which stresses the fundamental role of cognitive conflict in the construction of objectivity (Balacheff, 1988b). The epistemological solution is in principle rather simple: Knowledge construction consists in the progressive construction of mental representations, implicit or explicit, that are homomorphic to reality for some aspects and not for others. Representation is both active, pragmatic, and operational on the one hand, and discursive, theoretical, and symbolic on the other.

Objectivity is rarely complete, but in a sense there is always some objectivity in any representation. Quite often it is not the same representation that governs action and discourse.

The Theory of Conceptual Fields

The picture to be drawn is complex. This complexity comes mainly from the fact that mathematical concepts draw their meaning from a variety of situations and that each situation cannot usually be analyzed with the help of just one concept but rather requires several of them. This is the reason we have to study the learning and teaching of conceptual fields, that is, large sets of situations whose analysis and treatment require several kinds of concepts, procedures, and symbolic representations that are connected with one another. Examples of conceptual fields are additive structures, multiplicative structures, projective and Euclidean geometry, logic of classes, and elementary algebra.

The complexity of the picture comes also from the long-term development of mathematical concepts and procedures: For instance, it takes students many years to master additive structures. Some aspects of addition and subtraction are grasped by 4-year-olds, but there is one class of problems, requiring just one addition of whole numbers, that is unsuccessfully solved by the majority of 15-year-olds:

> Thierry has played two games of marbles. He has lost 17 marbles in the second game. He does not remember what happened in the first game. When he counts his marbles at the end, he finds that altogether he has won 25 marbles. What happened in the first game?

The reason for the failure lies mainly in the fact that transformations are adequately represented by directed numbers

$$x + (-17) = (+25)$$

and that the "true" operation is a subtraction of a negative number from a positive one. The addition $17 + 25$ is therefore counterintuitive. This difficulty concerns most problems that can be represented by the equation $a + x = b$ when b and a have a different sign.

Last but not least, the complexity of the picture comes from the fact that we need a theoretical framework that provides a strong articulation between problems to be solved and knowledge, and also between schemes, concepts, and symbols.

A valid body of knowledge on the psychology of mathematics education requires very systematic work, both theoretical and empirical:

- Analyze and classify the variety of situations in each conceptual field;
- Describe precisely the variety of behavior, procedures, and reasoning that students exhibit in dealing with each class of situations;
- Analyze mathematical competencies as organized schemes and identify clearly the invariant properties of situations on which the invariant properties of schemes rely (concepts-in-action and theorems-in-action);

- Analyze how language and other symbolic activities take place in such schemes, how they help students, and also how teachers use such symbolic intermediaries;
- Trace the transformation of implicit invariants, as ways to understand and act, into well-identified mathematical objects, which become progressively as real as physical reality and
- Trace the way by which students become conscious that procedures have a relationship of necessity both to the goals to be reached and to the initial conditions, and subsequently that theorems can be proved.

This is a program for research. We have only bits and pieces of information on these complementary lines of inquiry. Even for the conceptual field of additive structures, which has been fairly well studied, we need a lot more empirical research, especially with fourth and fifth graders and secondary school students.

It would require too much room to explain fruitfully a systematic classification of the situations in a given conceptual field and to provide empirical findings. Therefore we just trace the development of the concept of number as a product of the interaction of several categories of problems at different phases of the cognitive development of mathematical ideas.

The concept of number is a good illustration of the long-term process of knowledge acquisition. This concept addresses and solves some important problems that other concepts do not solve. It is our task to identify as clearly as possible the cognitive tasks that young children are tackling when they learn how to count and also the tasks that students are faced with at different levels of their development.

Essentially, young children make sense of the concept of number through situations of comparison and situations of combination. In such situations, they meet order and equivalence relationships on the one hand, addition and subtraction on the other. In other words, they need to know, and eventually to say, who has more (sweets, juice, cake, marbles, etc.), and also what the new state of a collection will be when one adds to it or takes from it some quantity, or one puts two parts together into a whole. Actually, comparison problems alone do not require the concept of number: Any order relationship and any ordered set of symbols (such as letters of the alphabet or ordered parts of the body) would be sufficient to compare discrete quantities and to rank a set of objects. It is addition and subtraction that give specificity to the concept of number: By associating a number to a collection (cardinal), children make it possible to add and subtract, and therefore to anticipate the expected value of a collection that is being increased or decreased. The same is true for the union of two parts into a whole.

The implicit theorem-in-action for this latter capacity is straightforward:

$$card\ (A \cup B) = card\ (A) + card\ (B),$$

provided $A \cap B = \emptyset$.

Comment: You can find the cardinal of the union of two collections by either counting it all [*card* $(A \cup B)$] or counting the two collections separately and then adding both cardinals [*card* (A) + *card* (B)].

Similar theorems must be implicitly recognized by children for them to be able to understand addition as an increase and subtraction as a decrease (Gelman & Gallistel, 1978) and to find the final state knowing the initial state and the transformation.

The recognition of these theorems goes together with the use of the counting-on procedure for addition (instead of the counting-all procedure), and the counting-down procedure for subtraction (for more details, see Bergeron & Herscovics, this volume).

Another interesting source of addition and subtraction problems for young children is the quantification of comparison relationships: *n* sweets more than, *n* sweets less than.

It is now well recognized that from these three basic relationships
- initial state/transformation/final state
- part/part/whole
- referee/comparison relationship/referent

can be derived many distinct classes of problems (Carpenter & Moser, 1982; Nesher, 1982; Riley, Greeno, & Heller, 1983; Vergnaud, 1982a).

But it is also necessary to recognize that, at the same time, children have a threefold experience of numbers: as measures (cardinals), as transformations, and as comparison relationships. If cardinals open the way to the concept of natural number, transformations and comparisons open the way to the concept of directed number: They can be inverted, and they can be combined in ways that are very different from the ways cardinals can be combined. For instance, one can trace the emergence of two interesting theorems-in-action concerning directed numbers.

At the age of 5 or 6, many children can recognize that an increase of *n* is cancelled by a decrease of *n*, whatever the initial state may be, at least for small numbers:

$$(+n) \circ (-n) = 0.$$

At the age of 8 or 9, most children can find the initial state, knowing the final state and the transformation, by inverting the transformation and applying this inverse transformation to the final state:

I = initial state, T = transformation, F = final state

Theorem-in-action: $F = T(I) \rightarrow I = T^{-1}(F)$.

The inversion of a comparison relationship (if John has *n* sweets more than Barbara, then Barbara has *n* sweets less than John) shows similar reasoning. There is some reliable evidence that one can introduce directed numbers as unary operations at the primary school level, represent them by symbols--for instance,

$$\begin{array}{ccccc} & +n & & -m & \\ \square & \rightarrow & \square & \rightarrow & \square \end{array}$$

--and operate syntactically on such prealgebraic sentences.

And yet, one must not forget that some problems with directed numbers are still difficult for most 15-year-olds, as we saw with the example above concerning Thierry's two games of marbles (page 23).

Actually, empirical results show that students can handle fairly well certain classes of problems involving transformations and relationships, whereas they fail on some others. This is the most important argument in favor of the conceptual field framework: There are different aspects and different operations for the same concept whose mastery takes several years, sometimes many.

The situation is again complex when children are faced with measures, transformations, and comparisons of continous magnitudes like length, weight, area, and volume. The handling of such magnitudes pushes children far beyond the concept of whole number in the direction of fractions (including decimal fractions) and even irrational numbers. The need for fractions smaller than 1 appears very quickly in the measurement of containers (half a glass), and teachers can also introduce situations that will eventually drive children to the need for very small numbers, such as finding the thickness of a sheet of paper (Brousseau, 1981). Another source of experience for decimal fractions is the task of approximating the measure of the side of a rectangle: What is the width of a rectangle whose area is 12 cm^2 and whose length is 7 cm? One can even ask primary school students to approximate the length of the side of a square whose area is 27 cm^2 (Douady, 1980).

The approach to decimals in this kind of teaching experiment relies mainly on a very important property of decimals: their density in the continuum of real numbers. Decimals enable us to approximate indefinitely any real number. The story of extension goes on with complex numbers, but there have been very few studies of the teaching and learning of complex numbers.

Let us stay for a while with the problem of negative numbers, as students have difficulties in working with them and interpreting them when they find such numbers appearing as solutions of equations or inequalities. It is interesting to recall that negative numbers were not accepted as numbers by mathematicians over a long period of history. With the advantage of historical hindsight, it is possible to identify the skills that Babylonian mathematicians had with first-order equations: They had a complete set of procedures for those equations that had positive solutions but classified those that would lead to negative numbers as unsolvable. Arabian mathematicians around the 10th century had arrived at a similar power over second-degree equations, recognizing that many had two positive solutions and hypothesizing that perhaps all should have two solutions. In time, this work led to a tentative acceptance of negative numbers, made stronger by the proposal of the notion of the number line whereby numbers could be seen to proceed in a negative as well as a positive direction. Similar considerations arose during the Renaissance, when the solutions to third-order equations were also categorized, solidifying the acceptance of negative numbers and giving impetus to the concept and eventual acceptance of another extension of the number concept, the "imaginary" numbers.

In fact, over 1500 years were needed for the rule of signs, first introduced by Diophantus, to become fully accepted by mathematicians. As Glaeser (1981) showed in a fascinating short historical study, several famous mathematicians like Stevin, Descartes, Maclaurin, Euler, d'Alembert, Carnot, Laplace, and Cauchy still had strange and partial conceptions of negative numbers because they either could not give any meaning to negative quantities; were unable to think in a homogeneous and coherent way about the number line, seeing it as two half lines rather than as a single totality; or were unable to reconcile the idea of an absolute zero beneath which nothing is supposed to exist with the idea of zero as an arbitrary origin. Glaeser identified some traces of these difficulties even in Cauchy's works: for instance, a confusion between the operative interpretation and the predicative interpretation of the "plus" and "minus" signs. How could we possibly imagine that students would easily understand negative numbers?

There are not many studies on the acquisition of negative numbers; they all reveal long-lasting obstacles in 15- and 16-year-old students, especially when they have to multiply a negative by a negative or when they come to a negative solution. Paradoxically, as we have seen above, there are some aspects of negative numbers that can be easily understood by primary school students: a negative transformation (decrease, loss, consumption, backward displacement), a negative relationship (less than, debt), or even a negative abscissa (below the ground floor).

This paradox can be solved at the theoretical level by the idea that negative numbers gain their meaning from different classes of problems that cannot all be mastered at the same level. Operations with negative numbers have a different meaning and a different power when they represent a decrease in a quantity, the cancellation or the compensation of a positive transformation, the inversion of a transformation or a relationship, the subtraction of two transformations, or the algebraic closure of the set of numbers for subtraction.

The paradox can also be solved at the empirical level by analyzing the variety of students' competencies in a variety of situations involving transformations, relationships, abscissas, directed numbers, and algebra, and by experimenting with original teaching experiments at the secondary and primary levels.

The concept of number is a striking example of the interconnection of many different aspects in the same concept.

First, these aspects are operational and related to different categories of cognitive tasks. Even for the same problem, different procedures may express different theorems-in-action: For instance, the knowledge is not the same that leads to inverting a transformation and applying it to the final state as that which leads to making a hypothesis about the initial state, applying the direct transformation to it, and correcting the hypothesis as a function of the result obtained.

The second important idea is that behavior is usually organized in schemes that can be used repeatedly in similar situations: Counting a set, executing a numerical operation like subtraction, or solving an equation of some type ($ax + b = c$, for instance) are all examples of schemes, at different levels of mathematics education, with different ingredients.

These schemes often imply that students work with linguistic signifiers or other symbolic representations: words in counting, digits and spatial display in subtraction, algebraic symbols and sentences in equation solving. But language and symbols also have the function of expressing concepts and theorems for communicating or for eventually generating a solution. By using words, symbols, or drawings of some kind, students identify relevant objects and relationships. It may also be the teacher who uses words, symbols, and drawings to help students. It may also be a peer. In all cases, the function of signifiers is to identify, select, and articulate information.

The problem of proof would not exist if we had no words or symbols to express well-formed sentences held as true. A proof is a relationship between sentences. But sentences are not mere declarations. They would be useless if they were not related to problems and to procedures. It is difficult to find examples of theorems that would have no connection to action and problem solving. As mathematics has become an enormous body of knowledge, one tends to see it as a set of propositions tied to each other in a deductive fashion. But the very origin of the feeling of necessity probably comes from the fact that what we do is determined by the aim we want to reach and the conditions under which we act. Proof is not discursive in essence but operational, and the first pragmatic proofs, as Balacheff (1988a, 1988b) calls them, come from two constraints: to act and to come to an agreement with others. Nevertheless well-defined words, formal symbols, and explicit syntax undoubtedly play a part in the progressive clarification of what a proof is. Here again one can see the contribution of language and symbols to thinking. Vygotsky comes to the aid of Piaget.

Some Educational Implications

Epistemological issues are almost totally overlooked by teachers, curriculum designers, and even researchers. And yet their importance for teaching and research can hardly be overestimated.

There is a gap between teachers' epistemologies and students' epistemologies, and that gap is reinforced by the fact that teachers usually do not question their own epistemologies or those implicit in textbooks.

The most commonly shared epistemology is the belief that mathematical activity consists in the discovery of timeless truths (Platonism), independent of culture, and that it is mainly a matter of logical reasoning. The specificity of mathematical conceptualizations and their relative character are rarely expressed, except in works like Lakatos's (1976) *Proofs and Refutations* and in some other pieces of research on the history of mathematics. Thus in education a major problem probably comes from the fact that many teachers take mathematical concepts as ready-made objects, without seeing that these concepts have to be built by students as functional tools that will enable them to deal with several kinds of situations.

The structural and descriptive vision of mathematics as it appears today is the result of a long history. Students have somehow to go through the same main conceptual difficulties, and they have to overcome the same epistemological obstacles that have been met by mathematicians. Mathematics is a "fallibilist social construction" (Ernest, 1985; Lerman, 1987), in the sense that concepts develop through negotiation with situations and with other people; concepts are culturally and temporally relative and potentially fallible. Objectivity appears in the shared public nature of theories and concepts, rather than through complete correspondence with the real world. Truth, proof, and rigor can also be seen to be relative. The relativist view roots knowledge in ideas, hypotheses, "bold conjectures," as Lakatos (1976) calls them, open to others to accept, change, or refute. This approach has rich possibilities for curriculum development and research in the teaching and learning of mathematics. Brousseau (1986b) has formalized this in his theory of didactic situations.

Another problem is that mathematics teachers have usually been good students of mathematics and have also been interested in the subject. As teachers, they have to teach a great variety of students whose goals, capacities, and interests are very different. To face such a situation, they must give up their personal view of mathematics, at least partially, and try to envision a wider spectrum of possibilities as to the meaning that mathematics may have for their students. The lack of consideration of the history and epistemology of mathematics in teacher training has certainly had bad consequences in teachers' vision of mathematics and mathematics education.

There is certainly not just one good perspective for the study of mathematics education. Psychology itself is one approach among several others, and psychology is not unique. There is much to be gained from theoretical and empirical studies emanating from different frameworks. But at least it is necessary that researchers, teachers, textbook authors, and curriculum designers recognize the importance of epistemological inquiry. By not doing so, they leave unquestioned important matters that actually condition their views and prevent them from raising key problems.

The introduction of some history and epistemology into the teaching of mathematics could be experimented with and reported on. It is worth studying whether the epistemology of mathematics contributes directly to the epistemology of mathematics education. A number of reports have been given of changes in teacher education to reflect these ideas, and further research should be carried out on teachers' attitudes and teaching styles and students' attitudes and learning styles.

A cultural and even ethnographic approach to mathematics teaching and learning, both outside school and in the classroom, is certainly very useful; it can reveal some astonishing discrepancies, like those found by Carraher (1988) between what is taught and what is used and between what we assume is meaningful to students and the ideas they actually work with. A cultural approach does not mean that mathematics is not good for all. On the contrary, it is becoming rapidly a universal problem that all students should know more mathematics and be confident with it. This problem is also cultural in the sense, perhaps, that different societies require different conceptions of mathematics and different conceptions of mathematics education.

Cognitive and developmental psychology are certainly essential in that they really question what a concept is; what an operational behavior is; how they develop; what part is played by action, perception, and language in concept formation; and what part is played by social interaction. We have sketched a synthetic theory of that process above. Emergence, extension, and refutation of conceptions exist in mathematics as well as in physics and biology. The fact that the distance between situations and models is shorter in mathematics must not hide the fact that one may view and solve differently the same problem and stick to wrong ideas for a long time.

It is essential that textbook authors include in their material some historical perspective and present a few important examples of change in mathematical ideas. It is also essential that students go through important changes in their own ideas by solving problems, discussing different conjectures and procedures, and becoming more conscious of their own conceptions and difficulties.

But mathematics education requires from teachers a better understanding of the interconnection of concepts, competencies, symbols, and situations in the long-term development of mathematical knowledge.

2 PSYCHOLOGICAL ASPECTS OF LEARNING EARLY ARITHMETIC

Jacques C. Bergeron
Nicolas Herscovics
with contributions by
Eric De Corte, Karen Fuson, James Hiebert, James Moser,
Pearla Nesher, Hermine Sinclair, and Lieven Verschaffel

Even in its prime, psychology showed an interest in early arithmetic, witness E. L. Thorndike's book *The Psychology of Arithmetic* published in 1922. Conversely, the teaching of arithmetic has always reflected the psychological theories in vogue at the time. Anyone over 50 remembers the endless hours spent in school reciting sums in unison to memorize them. We were then applying one of the principles of Thorndike's theory of associationism, the "law of exercise," thereby increasing the specific stimulus-response link (Mayer, 1977). Cognitive psychology today recognizes that higher mental processes are involved in the learning of early arithmetic, that is, the acquisition of the fundamental conceptual schemes of number and additive structures.

This contemporary view is expressed in the work of the International Group for the Psychology of Mathematics Education (PME). The need for more sophisticated theories can be illustrated by the notion of number, which fails to be described in terms of classical concept formation theory since it cannot be defined in terms of attributes or in terms of examples and nonexamples. Instead, number is now viewed as a conceptual scheme, that is, a network of related knowledge together with all the problem situations in which it can be used. Much of our research is also embedded in an epistemological framework, both in the general sense of the growth of knowledge and in the more restricted sense of genetic epistemology, which takes into account the development and maturation of the child. Underlying our philosophical outlook is the belief that mathematics is a way of thinking and as such cannot be merely "transmitted." This idea is at the heart of constructivism, which states that each individual must construct his or her mathematical knowledge or re-construct it in appropriate didactical situations.

In order to investigate the higher mental processes involved in the construction of early arithmetic, our research methods have borrowed heavily from psychology. The clinical methods of the Piagetians have enabled us to follow children's thinking through a dialogue by raising questions regarding their reasoning in the performance of a sequence of tasks. The translation of Soviet research in the psychology of learning and teaching mathematics (Kilpatrick & Wirszup, 1969) provided us with Vygotskian methods, Russian-style teaching experiments, enabling us to study "the very process of learning, as it takes place under the influence of pedagogy" (Menchinskaya, 1969, p. 89). All these research methods provide us with the possibility of identifying some of the cognitive obstacles encountered in the learning of arithmetic. Moreover, by carrying out some case studies, one can assess the cognitive potential of a pedagogical intervention.

Some Important Issues in Early Arithmetic

In the years since the founding of PME, some important research issues have been raised. These questions have been the subject of extensive research. An examination of studies dealing with the long-term evolution of students' arithmetical concepts shows that number and the different counting procedures play a much more important role in addition and subtraction, and for a much longer period, than expected (J. C. Bergeron, Herscovics, & Moser, 1984). For instance, in solving the missing addend problem "Cathy has nine pencils. How many more does she have to put with them so that she has fifteen pencils altogether?" 60% of all third graders use number facts, but the others still rely on counting procedures with about 30% *counting up from given* (reciting from 9 to 15 and counting the number words pronounced) and about 10% simply *adding on* (augmenting a set of 9 elements until it contains 15 and then counting the number of objects added) (Carpenter & Moser, 1984).

This example shows the need to pay greater attention to the various counting procedures used by children. But any study of counting must inevitably answer a more fundamental question regarding *what constitutes a number*. Such epistemological considerations are of great importance if we wish to understand the theories underlying research on the emergence of the number concept. Two opposing views can be identified: Piaget's earlier theory, which perceives logical reasoning as the basis for the construction of the number concept (Piaget & Szeminska, 1941), and another viewpoint which contends that numerical concepts evolve from the skills acquired through the quantification process. Questions arise regarding these two schools of thought. Are they really contradictory or might they not prove to be complementary in a broader perspective that considers number teleologically, that is, in terms of its uses?

Regardless of the theory one chooses, the teaching of early arithmetic still must contend with the young child's mind being affected by spatio-physical transformations.

For instance, for many kindergartners the elongation of a row will influence their perception of the quantity of objects present or the rank of a specific object in the row. The visibility of the objects has an even greater impact on most 6-year-olds since the great majority (75%) believe that hiding part of a row of chips affects the quantity, even if counting is involved. Of course, this does not mean that they fail to perceive the permanence of the objects, but it nevertheless leads them to believe that the quantity has changed (J. C. Bergeron & Herscovics, 1989). These considerations bring up the perenial question of maturation versus instruction. Should one wait until the children's logical thinking can overcome the effect of irrelevant figural transformations, or on the other hand, can instruction have any impact that would enable them to overcome their visual apprehension?

One of the first results of instruction, formal or informal, is the child's acquisition of the number-word sequence. Everyone agrees that its memorization cannot by itself be taken as an understanding of the number concept. Nevertheless, it is a prerequisite for the learning of the various counting procedures. Of course, one always knew that the child's knowledge of the number-word sequence increased with age. But the investigation of these numerical skills cannot be restricted to the study of the gradual extension of their range. As shown in the missing addend problem illustrated earlier, children use sophisticated procedures such as *counting up from given*. Surely, the development of such advanced counting procedures must rely on an equally sophisticated knowledge of the number-word sequence. Thus it is important to determine the various numerical skills related to the number-word sequence. These skills would represent a *qualitative* development in the child's mastery of the number words and would then raise the possibility of identifying different *levels* of cognition.

Discovering the different skills related to the number-word sequence opens up the possibility of experimenting with them in various cardinal and ordinal tasks. This possibility raises many interesting questions. For instance, will a child who knows how to recite the number words starting from a given number use this skill in counting a row of chips where the first six chips are hidden in front of him or her? More generally, will the acquisition of the various counting procedures help children in discovering the invariance of cardinality and ordinality under various irrelevant figural transformations? Most children in Western cities live in a numerically rich environment, and by the time they enter primary school they have acquired a fair amount of informal knowledge. How they formalize this numerical knowledge is little known. Does the teacher have to start from scratch, or do the children have some inkling about numerals and positional notation?

Children's counting skills greatly affect the choice of procedures available to them in the solution of the addition and subtraction problems they encounter in the first years of primary school. This is particularly true today since most countries have done away with the rote memorization of number facts. These observations raise many interesting new questions. If pupils' counting skills are limited to *counting all* (starting to count from 1) or *counting on* (continuing to count from a given number), can one expect them to handle addition and subtraction problems without *direct modeling,* that is, by using concrete objects or fingers to represent the different sets involved? Does the learning of more advanced counting procedures free them from using these objects, and does it enable them to solve a more complex class of problems?

Although the questions raised up to now have been about different counting and additive procedures, one should not infer that their mastery ought to be an end in itself. Indeed, procedural knowledge becomes valuable only if and when it can be used in meaningful problem situations. These problem situations need to be studied and analyzed from both a semantic and a syntactic perspective. The semantic analysis cannot be confined merely to the meaning of the words. It must look at a deeper meaning of the problem, that is, its "structure." Such an approach leads to important questions regarding the semantic categorization of additive word problems. Do all such problems involve the part/part/whole relationship? Can one distinguish between static and dynamic structures? Can the semantic categories be defined in terms of basic cognitive schemes? Do they represent distinct cognitive levels?

Of course, the classification of additive problems and the hierarchy it provides cannot be dissociated from the additive procedures used by children. But the procedures they use may also vary according to nonsemantic factors. This variation leads to the likely hypothesis that syntactic factors such as the relative size of the numbers involved and the order of presentation of the given sets could affect the procedures used to solve the problems. In fact, would the choice of procedures not vary simultaneously according to both the semantic and syntactic contexts?

In these few paragraphs we have raised some of the main issues on which research in early arithmetic has focused the last 15 years. Results in this field have proved to be quite fruitful. The investigations have contributed significantly to our understanding of the cognitive processes involved. We report below on some of the studies that have dealt with these questions.

Theoretical Approaches to Number Acquisition

As noted above, two major competing positions emerged early in the recent era of number research. One was Piaget's (1952) theory that emphasizes the primacy of logical-reasoning abilities in the development of number concepts and skills. A very different view was proposed by a number of researchers. This view, perhaps presented most comprehensively by Klahr and Wallace (1976), suggests that number develops through the acquisition of several separate quantification skills.

To understand Piaget's (1952) view of number concept development, it is useful to review a critical distinction Piaget (1970) makes between logico-mathematical knowledge and physical knowledge. The first type of knowledge is generated by internal mental processes, whereas in the second, the individual has to start by taking into account the physical properties of the objects in the environment as he or she perceives them. The first arises from deduction and is verifiable by logical reasoning;

the second arises from induction and is verifiable by empirical test. Piaget viewed number as a logico-mathematical concept that is *constructed* by the child rather than a physical concept that is discovered through sensory perceptions. An understanding of number requires a prior understanding of key logical concepts such as conservation, class inclusion, and seriation. Although Piaget acknowledged that certain quantifying skills, such as counting, are acquired prior to the full development of these logical concepts, he contended that they take on meaning only through the application of these concepts.

The theories that propose that children acquire number through deploying quantification skills are in contrast to the theoretical positions of Piaget (Klahr & Wallace, 1976; Schaeffer, Eggleston, & Scott, 1974; Young & McPherson, 1976). Klahr and Wallace postulated three distinct quantification processes: subitizing, counting, and estimating. The function of these processes is to generate quantity or numerosity "symbols" of sets for mental manipulation. These processes or skills are hypothesized to develop in an invariant sequence. Subitizing (the instant recognition of the number associated with a configuration) is the first skill to be acquired and is part of the basis on which the child's understanding of number develops. Subitizing also plays a vital role in the later development of counting and estimating. These latter two skills develop concurrently, but since numerical estimation requires the acquisition of several additional component skills, it reaches maturity later than counting.

A number of theories that influenced the work in early number do not fit well in either the logical reasoning or the quantification skill camps. Two of the most widely cited "hybrid" theories are those of Ginsburg and Gelman.

Ginsburg (1975, 1976, 1977, 1982), like Piaget, believed that the number concept cannot reach completion without certain logical-reasoning abilities. Ginsburg suggested that the development of preschool children's knowledge of number concepts can be portrayed as a progression through two cognitive systems. System 1 is informal in that it develops outside of formal school instruction, and it is natural because it does not depend on social transmission or specific cultural experiences. Children who are operating within this cognitive system are able to discriminate between numerosities in terms of "more" and "less," using well-developed perceptual skills. System 2, like System 1, is informal, that is, it develops prior to formal instruction. However, System 2 is not a natural system because it depends upon socially transmitted knowledge. Counting is the primary characteristic of System 2 and provides the child with a widely applicable and reliable quantification skill.

Gelman (1972, 1977, 1982; Gelman & Gallistel, 1978) dealt with the contrast between logical reasoning and quantification in a somewhat different way. She began by emphasizing the distinction between processes of quantification and processes of reasoning. Then she distinguished between reasoning about specified numerosities (collections that have been quantified) and unspecified numerosities. Number, according to Gelman, should be discussed only within the context of numerosities that can be accurately represented. Counting was believed to be the basic and primary quantification skill. It serves to reliably determine the numerosity of sets and thereby defines the domain within which children first learn to operate with number. The development of the counting skill over the preschool years is guided by the presence of five counting principles that define a successful counting procedure. The principles are believed to form a scheme in the Piagetian sense. A unique aspect of Gelman's theory is the conjecture that these principles are "wired in" and unfold with development. Thus, the principles are believed to precede acquisition of the related skill so that children's behavior is rule-governed rather than capricious. In other words, young children possess counting principles in search of appropriate skills.

Both Ginsburg's work and Gelman's reflect the shortcomings of information-processing theory and the Piagetian dichotomy between logico-mathematical knowledge and physical knowledge. Information-processing is limited to a study of the procedures it can simulate and cannot account for much of the informal mathematics gathered by young children. On the other hand, the Piagetian dichotomy results from a confusion of logic with mathematics. Whereas logic can be described by mathematical models, Russell and Whitehead's attempt to reduce mathematics to logic did not fare very well (Davis & Hersh, 1981). As an alternative, we suggest a distinction between *logico-physical* understanding, which results from thinking about either procedures applied to physical objects or spatio-physical transformations of these objects, and *logico-mathematical* understanding, which results from thinking applied to either procedures or transformations dealing with mathematical objects (Herscovics & Bergeron, 1988a). In this framework, one can contend with the reflective abstraction of actions operating in the physical realm without necessarily describing it as somehow being mathematical. For instance, a child establishing a one-to-one correspondence between two sets of objects could be considered as providing evidence of procedural understanding of a logico-physical nature, whereas such a correspondence between a set of objects and the number-word sequence could be qualified as procedural understanding of a logico-mathematical nature. Such a distinction answers the need expressed by Steffe and his colleagues (Steffe, von Glasersfeld, Richards, & Cobb, 1983) to distinguish between a physical unit and an arithmetic unit.

The distinction between the physical world and the mathematical world provides us with the opportunity to define number teleologically, that is, in terms of its functions and uses. Initially, numbers are used to answer two distinct questions: "How many objects are there in a given set?" and "What is the rank of an object in an ordered set?" But well before children have any knowledge of number, they can distinguish between one and several objects. Their ability to perceive several physical units is all that is needed for their conception of *plurality*. In this sense, number in its cardinal function can be viewed as a *measure of plurality*. Similarly, children are aware of the *position* of an object in an ordered set even without being able to determine its rank numerically. Hence in its ordinal sense, number as rank can be viewed as a *measure of position* in an ordered set. Our definition of number answers the need to distinguish between what Gelman (1972) calls "numerosities" (collections that have been quantified) and unspecified numerosities.

From a teleological perspective, the apparent contradictions between the two schools of thought dissolve, and instead their complementarity becomes apparent. For instance, everyone is familiar with Piaget's test on the so-called "conservation of number" in which a child has to decide whether two rows of objects have the same quantity after one of the rows is elongated. A more appropriate name for this task would be "conservation of plurality" since no enumeration is involved. This was recognized by Piaget's collaborator Gréco (1962), who included counting and defined as the *conservation of quotity* the child's ability to predict the number of objects in the elongated row after having enumerated the other row. Gréco found that three quarters of the twenty 5-year-olds he had interviewed did not believe that the two rows had the same quantity even after counting. Thus their quantification seemed somewhat meaningless. Both the conservation of plurality and the conservation of quotity are essential in the child's abstraction of the cardinal aspect of number (Herscovics & Bergeron, 1989). Of course, a similar argument holds for the ordinal aspect. The complementarity of the logico-physical and the logico-mathematical processes is quite evident.

Developmental Considerations

Numerical constructions seem to have deep biological roots. The newborn can make certain "numerical" discriminations both in the visual presentation of dot patterns and in the auditive presentation of rhythmical sequences. It is possible to observe, from the age of 12 months or so, manifestly intentional activities that without doubt have to do with plurality. Admittedly, it is difficult to construct a theory of number development that includes such presumably built-in capacities as well as mathematical knowledge.

Infants can spend considerable time putting objects one by one into a container and taking them out again. Twenty-four-month-old babies spontaneously construct one-to-one correspondences by putting one object into one container, the next into another, and so on. These activities are at the same time the source for later logico-mathematical operations and for spatio-physical operations. The initial lack of dissociation between spatio-physical activities and logico-mathematical thinking continues for some years. At first, the child cannot dissociate the plurality of a collection of objects from the space in which they are contained: If the objects occupy a larger space, the child thinks that there are more.

If number construction is indeed a process without clear end or beginning, are there nonetheless certain milestones along the road towards ever more complex numerical systems? Educational practice is implicitly built on a supposed milestone: Around the age of 6, children are supposed to be ready for "formal" arithmetical lessons. A psychological milestone is also reached at about this age when children succeed on the well-known conservation of plurality task (Piaget & Szeminska, 1941). According to Piaget, "the best criterion for the appearance of mental operations . . . is the constitution of invariants or notions of conservation" (Piaget & Inhelder, 1963, p. 119). In this perspective, success in the plurality conservation task is proof of a system having been constructed that leads to some necessary deductions: Verification by counting or rearranging the objects has become nonpertinent (the deductive reasoning is not contradicted by the physical evidence).

Arithmetical reasoning, even of the simplest kind, can exist only when some sort of coherent system allowing for deductions has been constructed. Yet it is certainly not true that the reasoning system manifested by success in the conservation task marks a total divide between being able to deal with addition, subtraction, multiplication, and division and not being able to do so. It does not give immediate access to, say, the program of the first three grades, nor in contrast is it true that before this achievement no coherent numerical reasoning is at all possible. Gréco (1962, chap. 1) and many researchers more recently have shown the importance of counting as an action that favors the construction of a reasoning system and allows for certain deductions. Certainly, educators and parents know that 4- to 5-year-olds are already capable of coherent numerical reasoning in some circumstances but seem to have peculiar difficulties in others. If it were possible to clarify what makes some types of problems accessible to preschoolers and others inaccessible to grade school students, education could certainly profit.

Recent studies have tried to answer some of these questions. J. C. Bergeron and Herscovics (1988; Herscovics & Bergeron, 1988b) have investigated the kindergartner's understanding of plurality and position. Both notions can be considered part of Ginsburg's (1976) System 1 type of informal knowledge since no enumeration is involved in the suggested tasks. By the time they are 5 to 6 years old, virtually all children can use visual estimation successfully to compare two sets and decide which one has more, which one has less, where there are many or few, and whether the two sets have the same plurality. They can also decide whether an object in a row comes before or after another one, which one is first or last, whether an object is between two other ones, and whether a toy horse along a row of horses is together with or coming along at the same time as another horse. What is more important, practically all children in this age group can go beyond mere recognition of these notions and use a one-to-one correspondence to generate sets subject to the above plurality and positional constraints.

Regarding the kindergartner's perception of the invariance of plurality with respect to irrelevant spatio-physical transformations, it seems that maturation has a marked effect, as shown by comparing groups of kindergartners of average age 5:8 and 6:2. For instance, when randomly disposed cubes are dispersed, the percentage of children who believe that the plurality of the set is affected ranges from 43% for the younger ones to 22% for the older ones. A simple displacement of each cube without any change of the space occupied by the set also is perceived as having an effect by about 30% to 12% of the children according to age; between 27% and 6% in this age range believe that plurality has been affected by a change of perspective obtained by rotating a plate of cubes. The percentage of children who believe that the plurality of a single row is changed after it has been stretched out varies from 45% for the younger ones to 19% for the older ones. No wonder then that the success rate on the Piagetian conservation of plurality test ranges from 24% to 65% in this age bracket. But it is the visibility of the objects that most affects the kindergartner's perception of plurality. When a single row of chips is inserted into a partially opaque plastic bag so that three of the chips are no longer visible, between 91% and 72% of the children think that the quantity of chips has changed. When two such rows are compared, one completely visible, the other one partially hidden, nearly all (between 97% and 87%) believe that the two rows no longer have the same plurality (J. C. Bergeron & Herscovics, 1988; Herscovics & Bergeron, 1988b).

The kindergartner's perception of the invariance of the rank of an object in a row when the row has been elongated varies also with age. Among those closer to 5 years, 62% believe that the rank has changed, in contrast to 37% of the 6-year-olds. On the other hand, visibility of the objects seems to have a smaller impact on rank than on plurality. When in a row of images of trucks, the first three are hidden under a "tunnel," 79% of the younger children, as compared with 53% of the older ones, believe that the position of the remaining trucks in the row has changed. And yet, most of these children know that the rank of an object in a row depends on the quantity of objects preceding it. For example, when the first car in a row is removed, between 80% and 91% of the kindergartners indicate that it affects the rank of the remaining objects. Visual perception also plays a preponderant role when two rows of objects are present. Children can easily use a one-to-one correspondence to establish that two cars have the same position in two rows next to each other. However, if the two rows are translated by different lengths so that the two selected cars are no longer side by side, most children (90% of the younger ones and 81% of the 6-year-olds) do not believe that both still have the same rank.

These results confirm that the dissociation between spatio-physical activities and logico-physical deductions based on visual perceptions is still quite present in children on the verge of primary school. This dissociation must be taken into account during the first years of formal arithmetic instruction.

The Structuring of Numerical Knowledge

The child's ability to count is inevitably based on the acquisition of the number-word sequence. Most middle-class English-speaking children below age 3½ are working on learning the number-word sequence to ten, and most children between 3½ and 4½ are working on the words between ten and twenty. A substantial proportion of children between 4½ and 6 are still imperfect on the upper teens (the words between fourteen and twenty), but many are working on the decades from twenty through seventy (Fuson, Richards, & Briars, 1982). About half of first graders and almost all second and third graders know the sequence to one hundred and are working on the sequence above one hundred. Of course, children's ability to recite the correct sequence of number words is very strongly affected by their opportunity to learn and to practice this sequence. Children within a given age group show considerable variability in the length of the correct sequence they can produce.

The incorrect sequences produced by children before they have learned the standard sequence have a characteristic structure (Fuson et al., 1982; summarized in Fuson, 1988). Crucially important number-word sequence learning continues long after the child is able to produce the number words correctly. This continued learning manifests itself in an orderly succession of new abilities. Groups of these abilities require a representation of the number-word sequence that differs *qualitatively* from the representation of the sequence at other levels; these have been designated as belonging to five different levels (Fuson et al., 1982). This elaboration of the number-word sequence is a lengthy process ranging at least from age 4 to age 7 or 8. The elaboration occurs first at the beginning of the sequence; different parts of the sequence can be at different elaborative levels at the same time.

The five levels of elaboration are (a) *string level*, the words are a forward-directed, connected, undifferentiated whole; (b) *unbreakable list level*, the words are separated, but the sequence exists in a forward-directed recitation form and can only be produced by starting at the beginning; (c) *breakable chain level*, parts of the chain can be produced starting from arbitrary entry points rather than always starting at the beginning; (d) *numerable chain level*, the words are abstracted still further and become units in the numerical sense, and thus sets of sequence words can themselves represent a numerical situation and can be counted or matched; and (e) *bidirectional chain level*, words can be produced easily and flexibly in either direction (Fuson, 1988; Fuson et al., 1982). These different levels are marked by increasingly complex sequence abilities: becoming able to start and to stop counting at arbitrary number words, to count up a given number of words, and to count backwards starting and stopping at arbitrary number words or counting down a given number of words. Children also increase their ability across these levels to comprehend and to produce order relations on the words in the sequence.

The structure of a particular number-word sequence seems to affect the kinds of errors children make in learning that sequence (Fuson, 1988). Thus, the above results with respect to the acquisition of the English sequence of number words may apply only to sequences that share its features. However, the developmental representational levels involved in the elaboration of the number-word sequence depend less on particular features of the sequence and consequently seem to apply much more widely (Davydov & Andronov, 1981; Saxe, 1982). Instructional practices may also affect the manifestation and perhaps even the development of these levels (e.g., Hatano, 1982).

Counting objects distributed in space requires the person counting to establish a one-to-one correspondence between the objects and counting words. Because objects are located spatially and words are located temporally, some sort of intermediary is needed to connect the two. The counter establishes this correspondence by using an indicating act (some variation of pointing or of moving objects) that has both a temporal and a spatial location (Fuson, 1988). The indicating act thus serves as a mediator and itself involves two correspondences: the correspondence in time between a word and the indicating act and the correspondence in space between the indicating act and an object. The most common kinds of correspondence errors 3- through 5-year-old children actually make in counting objects *in a row* are essentially of five different types. Two of these violate the word-point correspondence: In *point-no word* errors children point without saying a word, and in *multiple words-one point* errors they say two or more words while pointing once at an object. Two other common errors violate the point-object correspondence: In *object skipped* errors an object is skipped over without being counted, and in *multiple count* errors an object is counted and then immediately counted again. The fifth common error violates both correspondences: In *multiple points-one word* errors an object is pointed at two or more times while one word is said.

The errors listed above occur in counting a row of objects. When the objects are laid out randomly, partitioning the elements or remembering which objects have been counted are other sources of errors. Gelman and Gallistel (1978) have shown that among 5-year-olds, 23% of the counts are incorrect because of partitioning errors and another 14% are wrong because the child has missed or double-counted the last item. With 6-year-old kindergartners, Herscovics, Bergeron, & Bergeron (1986) have found three distinct procedures: *visual counting* (V), in which the child keeps track of the objects visually without any physical contact; *touch-counting* or tagging (T), where each object is touched without any displacement; and *physical partitioning* (P), where the child separates the objects as they are counted by pushing them aside one at a time. In the first two procedures, the partition is established mentally, and this leads to most of the counting errors (about 20%) found at that age.

In their investigation of children's counting, Steffe et al. (1983) have carried out some case studies on the effect of the visibility of the objects to be enumerated by hiding some elements in a row of chips. A. Bergeron, Herscovics, & J. C. Bergeron (1986) tested this task on a larger sample of 6-year-olds ($n = 23$) who were told that 6 chips were being covered by a small cardboard and were asked to find how many chips were glued on the whole cardboard. The most popular solution procedure was *figural counting* (39%), that is, counting from 1 to 6 while moving one's finger over the screen covering the hidden chips and then counting the visible ones. Many children using this procedure failed in their verbal count over the hidden part although all remembered that 6 chips were covered. The next most common procedure was *counting on* from 6 (26%); the low popularity of this procedure is surprising in view of the fact that 22 of the 23 children could recite on from 6 in the number-word sequence. Two children used *double counting* by counting first the 5 visible chips and then 6 more, keeping track on their fingers. What is remarkable here is the ingenuity shown by most of these kindergartners who, when faced with a task they had never handled before, could nevertheless spontaneously come up with sophisticated counting procedures.

Yet, one cannot assume that these children already perceive the full implications of the enumeration procedure. Piaget himself (1973) as well as Gelman and Gallistel (1978) observed that after counting a given set, some children do not realize that a change in configuration or a change in the order in which the objects are chosen in the count will still result in the same cardinality. Ginsburg (1977) also pointed out that some young children can enumerate a given set several times and obtain different results without inducing any cognitive conflict that might bring about a sense of contradiction. The extent to which the invariance of cardinality is perceived by children on the verge of entering primary schools is important to the teacher. In separate investigations involving four groups of kindergartners (22 $\leq$ n $\leq$ 36), Herscovics and Bergeron (1989; Herscovics et al., 1986) found that between the ages of 5 and 6½, most kindergartners (from 86% to 91%) believe that the question "how many" can have *only one correct answer*. Regarding the invariance of the cardinality of a row with respect to the *direction of enumeration,* most of them (between 69% and 91%) no longer feel the need to count when asked how many they would find if the direction of their original count was reversed. In the case of a randomly displayed set which they have counted, after the objects have been *dispersed* but remain in full view, 63% feel the need to count them again when asked how many there are. However, hiding the set after it has been dispersed yields a better performance with a higher percentage realizing that the quotity has not changed (between 86% and 91%). Clearly, the visual perception of the dispersed set still affects about 25% of the children.

The kindergartner's informal knowledge extends beyond enumeration. Virtually all children in this age group living in urban middle-class environments can recognize and generate numerals meaningfully. The spontaneous written representation of numbers by young children has been investigated in different countries (Allardice, 1977; Sastre & Moreno, 1976; A. Sinclair, Siegriest, & H. Sinclair, 1983). More recently, J. C. Bergeron, Herscovics, and A. Bergeron (1986) asked three groups of kindergartners (27 $\leq$ n $\leq$ 32) "to leave a message for a friend that would tell him how many little cardboard houses were being shown (7)." Most children (92% and up) between 5 and 6½ could leave an appropriate message. A majority ranging from 51% to 74% simply wrote "7," thereby assigning it its cardinal meaning. The next most common form of representation was the sequence "1, 2, 3, 4, 5, 6, 7." Between 15% and 27% of the children left these as a message, thereby indicating that they still maintained a one-to-one correspondence between the set of numerals and the set of objects, just as in counting. The representation by seven drawings or seven tally marks proved to be relatively rare (8% and 3%).

But the kindergartner's knowledge of numerical symbolization is even more extensive than revealed in the above numerical messages. Asked to write out all the numbers they knew, between 48% and 87% could write numbers beyond 9. The results indicate that well before receiving any formal instruction, children have some grasp of *positional notation*, that is, the appropriate concatenation of digits and their global perception (12 is no longer seen as "one and two" but as twelve). In the child's acquisition of positional notation three distinct levels of understanding can be found. At the first level, that of *juxtaposition*, the child is aware that the digits are written side by side but their relative position is not yet viewed as important. A second level, labelled *chronological*, is revealed when children are asked to write from right to left: In wanting to write twelve they will put down 1 and then 2 to the left of it, thus obtaining 21. They are so concentrated on the order of production that they ignore the end product. A third stage, the *conventional* level, is achieved by the child who is able to write two-digit numbers in such a way that the conventional order prevails even when writing from right to left.

As can be seen from the knowledge we have gathered about the child's acquisition of number, research in this field has proved quite fruitful. And its importance becomes evident when we consider the next conceptual scheme, that of additive structures, which refers to the notions of early addition and subtraction.

Additive Structures: Children's Solution Procedures

The procedures employed by younger students as they attempt to solve one-step verbal problems in addition and subtraction may be categorized into three main types: *direct modeling* with physical objects, *verbal counting*, and *mental strategies* involving direct recall of some basic addition or subtraction fact. When numbers are sufficiently large, children will use traditional algorithms taught in class. However, a discussion of the use of algorithms is beyond the scope of this chapter. The three categories may be thought of as developmental stages that progress in level of abstraction. Depending upon the problem type and upon the size of numbers contained in the problem statement, individual children have been observed operating at different levels of abstraction. Thus, there is no clearcut division between stages (Carpenter & Moser, 1984).

Direct Modeling

In this category the child begins with the construction of one or more sets of visible objects. In addition to familiar manipulative materials or fingers, the objects may include written tally marks or some pre-existing collection of objects in a room such as floor tiles or chairs, in which case the "construction" consists of the visual delimitation of a subset of that collection. The distinguishing characteristic is that the objects are used as direct representations of the problem entities, and the actions performed on those objects are seen as representations of the actions or relationships contained in the verbal problem setting.

After initial construction of one set:

- *Incrementing.* The initial set is incremented one object at a time with that action terminating when a specified number of objects has been adjoined to the first set. Counting the number of objects in the total collection provides the answer to an addition problem.

- *Adding-on.* The initial set is added-on to one object at a time. That action terminates when the total set has reached a specified size. Counting the number of objects added to the initial set provides the answer to a subtraction problem.

- *Separating from.* A subset of a specified size is removed or separated from the initial set. Counting the number of objects in the remainder set provides the answer to a subtraction problem.

- *Separating to.* Objects are removed one at a time until the remainder set is of a specified size. Counting the number of objects removed provides the answer to a subtraction problem.

After initial construction of two sets:

- *Counting-all.* The child counts the union of the two sets to determine the answer to an addition problem. Several variations have been observed in the way children form the union of the two original sets.

  - *Unary.* One set is kept stationary in its originally constructed position and the other is then moved from its initial position to a new position immediately adjacent to the first.

  - *Binary.* Both sets are moved together to form the union.

  - *Stationary.* Neither set is moved and the act of union is made visually or mentally.

- *Matching.* The two sets are put into a one-to-one correspondence. Determining the size of the excess of the larger set over the smaller provides the answer to a subtraction problem. This determination may occur in two ways.

  - *Take-away.* The part of the larger set that is unmatched is separated physically or visually from that set and then counted.

  - *Add-on.* After the initial one-to-one correspondence is made, a new third set of previously unused objects is added on to the smaller of the two original sets until it is equal in number to the larger set. Counting the added on objects determines the answer.

The relatively rare procedure involving the initial construction of three sets has been documented by De Corte and Verschaffel (1987a). Depending upon the size of the sets constructed and upon whether objects or fingers are used as models, different techniques for constructing and counting sets may be employed (Baroody, 1987; Siegler & Shrager, 1984). The most basic method is to utter the counting words "one," "two," and so on. But when the size of a set is relatively small, subitizing may be used. When fingers are used, children may construct models using idiosyncratic kinesthetic or patterning methods. In the more advanced stages, when children are in transition to the next level of abstraction, the counting of the final union may not need to begin with "one" but may, instead, begin with the cardinal count of one of the two original sets.

Verbal Counting

Procedures in this category are identified by the entry into the sequence of counting words at a point other than "one" followed by forward or backward counting that ends when some rule is applied. One of the requirements of these procedures is the ability to execute a double count (usually simultaneously) such that counting words themselves are counted. This double counting, or tracking process, may be assisted by the use of objects or fingers. However, the objects here do not stand for the given problem entities; rather, they represent counting words (see Fuson, 1982).

- *Counting forward from.* Forward counting from the initial entry point is terminated when a specified number of counting words has been uttered. The final counting word spoken provides the answer to an addition problem. The entry point may correspond either to the first-named addend, the second-named addend, the smaller (MIN) addend, or the larger (MAX) addend (Groen & Parkman, 1972). This is a counting analogue to the incrementing procedure described earlier. It is also known by the name of counting on (Carpenter & Moser, 1984; Fuson, 1982).

- *Counting forward to.* Forward counting from the initial entry point is terminated when a specified number word is uttered. Counting the number of words spoken determines the answer to a subtraction problem. This is the counting analogue to adding-on. This procedure is also known as counting up from given (Carpenter & Moser, 1984; De Corte & Verschaffel, 1987a).

- *Counting backward from.* Backward counting from the initial entry point is terminated when a specified number of counting words has been uttered. The final counting word spoken (or its successor in the backward string) provides the answer to a subtraction problem.

- *Counting backward to.* Backward counting from the initial entry point is terminated when a specified number word is uttered. Counting the number of counting words spoken determines the answer to a subtraction problem.

When utilizing these counting procedures, many children operate at a subvocal level. The evidence for the incidence of these procedures does not always come from direct observation, but results from postsolution questioning by an interviewer.

Mental Strategies

These procedures fall into two major categories: recall of known basic addition or subtraction facts and so-called derived facts or "heuristics" (Carraher, Carraher, & Schliemann, 1987).

- *Direct recall.* Either through formal schooling or through contact with parents and siblings, children come to know certain basic facts that can be recalled from memory without any obvious recourse to counting. When recalling a direct fact needed to solve an addition problem, children may use the fact that relates directly to the numbers in the order presented in the verbal problem, or they may commute the order and begin with the larger number first. For example, if the number order is 5, 8, some may say, "Five plus eight is thirteen," while others may say, "Eight plus five is thirteen" (De Corte & Verschaffel, 1987a). For subtraction, some may recall a direct subtraction fact. But, perhaps of more interest when analyzing behavior relative to certain semantic variations, many children will recall the related addition fact. For example, if the presented numbers are 12 and 5, the child might say, "Five plus seven make twelve, so the answer is seven."

- *Derived fact.* Many of the procedures in this category involve decomposition of one of the given numbers into smaller parts so that one of the parts can be used together with the other given number as a known fact. The known facts that typically come into play are doubles (e.g., 6 + 6) and combinations of 10 (e.g., 7 + 3). A less frequent but equally interesting instance of a derived fact strategy involves what is often called compensation. A number that is 1 or 2 larger (or smaller) than a given number is used as part of a basic fact, and then adjustment is made to the final answer or to the other number.

The range of procedures used by children in the solution of addition and subtraction problems is obviously quite extensive. Yet all of them come into play and are selected on the basis of the structure of the problem presented.

The Semantic Aspect of Additive Word Problems

One of the more surprising findings in the last decade of research on additive word problems in arithmetic is that word problems that have the same logical structure and call for the same mathematical operation are nevertheless handled quite differently by young children (Nesher, 1982). Nesher found, for example, that a subtraction word problem such as "Joseph gave Dan 10 dollars. How many dollars are left, if Dan spends 3?" is very easy. The following percentages of success were obtained for such problems in four separate schools: 94% (n = 967), 89% (n = 22), 89% (n = 256), and 85% (n = 287). However, for the following subtraction word problem, the level of success was much lower: "Joseph and Roland had 7 marbles altogether. Three of them were Joseph's. How many of them were Roland's?" In this case the percentage of success in the same four schools was 52%, 46%, 49%, and 41%. One can therefore note that the kind of operation to be performed as a solution to the word problem is not the significant factor but rather that other variables are needed to explain this phenomenon.

In the last decade researchers from many countries (Belgium, France, Israel, and the USA) have repeatedly noted the same well-defined regularities and have tried to find out why they occur (Carpenter, Hiebert, & Moser, 1981; Carpenter & Moser, 1984; Carpenter, Moser, & Romberg, 1982; De Corte & Verschaffel, 1985, 1987b; Nesher & Katriel, 1977, 1978; Nesher & Teubal, 1975; Riley, Greeno, & Heller, 1983; Vergnaud, 1982a). It was first found that additive word problems can be categorized into three main semantic categories: *combine* problems, which are static in nature; *change* problems, which are dynamic in nature, describe a kind of on-going event, and are clearly sequential in time; and a *compare* type of additive problems that compare a quantity that one has to a given referent.

The three semantic categories have been explained in terms of the basic cognitive schemes that underlie them, and they represent cognitive developmental levels. At the first level, which occurs around the age of 5 or 6, the child is already able to handle single sets and has the knowledge needed to count them. This knowledge is sufficient to solve simple change problems, in which the "unknown" is at the end of the sequence. In this case the child relates to each string of the text separately; he or she acts accordingly but cannot relate to more than one set at a time. This procedure, however, is not sufficient to solve other kinds of problems, including the second example above. In that example, the child has to refer to the same set using two different descriptions (the marbles that Roland and Joseph have together and the marbles that each of them has separately). Nor can children at this age solve a compare problem that requires comparing two different sets simultaneously. They cannot rearrange a text that is not ordered according to the solution path since they would have to consider all the strings of information concurrently as well as reorder them.

At the second level, children are able to relate a change that occurred in the initial set to the relevant action in a more causal manner. They are able to estimate the direction of the change (increase vs. decrease). They also know the arithmetical operations that describe the directions of this change. This knowledge enables children to solve problems in which the question is directed to the change in the quantity caused by the action described in the text. The following problem is an example: "Jim had 5 stamps and he bought some more. Now he has 8 stamps. How many stamps did Jim buy?" (a so-called Change 3 problem). In order to solve this problem, it is not sufficient that children understand each string separately. They need to understand the connection between the first and the third strings--that the relation expressed in the text is of an "increase" that corresponds to the temporal order of the events. Their association of "increase" with the + operation will then lead them to perform the mathematical operation: $5 + \square = 8$. At this level children are able to solve this problem numerically by "counting on" from 5 until they reach 8.

At the third level, children have already acquired a part/part/whole scheme that enables them to handle a static situation in which they themselves have to impose a structure on the situation described by the verbal text. Combine 2 (subtraction) problems or Change 5 and Change 6 problems can then be solved (around the age of

7 or 8). For example, a child may be given a problem such as "John has some stamps. He bought 5 more stamps. Now he has 8 stamps altogether. How many stamps did John have to start with?" (Change 5 problem). This kind of problem could not be solved at an earlier level. The initial quantity is not known and, therefore, the child cannot solve by counting on. At Level 3, however, children are equipped with a part/whole scheme that enables them to ignore the temporal order of the events described in the text and relate to the information embedded in the text according to the role it plays in the part/whole scheme. They can now perform the arithmetic operation that is needed for finding a missing part, which is the subtraction operation. It should be noted that the nontemporal scheme of the part/whole relationship enables children to perform a subtraction operation in spite of the fact that the text describes an "increase" situation (he bought and thus now has more than in the beginning).

Finally, but not before the age of 9 or 10, the child is able to handle two-place relation schemes involving the comparison of two disjoint sets. These are needed to solve all kinds of compare problems, as for example: "Dan has 5 marbles and this is 4 marbles less than John has. How many marbles does John have?" (Compare 6 problem). Here the child has to handle a two-place situation--*less* (Dan's marbles, John's marbles)--and perform an addition operation in spite of the word *less*.

Thus, one has to understand the growth of the child's ability to solve word problems not as an isolated numerical issue but rather as part of a more global cognitive development within which the child's schemes of action are changing, enabling him or her to comprehend and solve more complicated situations.

In the extensive research on children solving additive word problems, many other structural variables that contribute to the child's level of performance have been identified. Some of these variables are semantic, as for instance: (a) the presence of some lexical items that serve as cue words for certain operations (such as *more* for addition and *less* for subtraction); and (b) the child's acquaintance with the described situation, that is, the context. Some of the variables are syntactic, as for example: (a) the location of the string of the information that describes the missing number (the actual number about which the question is raised, whether it is the first, second, or third number in the equation that is missing); and (b) the relation between the order in the text and the order of the events described in the text (e.g., "John now has 8 marbles after receiving 3 marbles from his brother. How many marbles did he have to start with?"). In this problem the order of events contradicts the order of the information given in the text.

We are now in a situation where it is possible to arrange additive word problems in some order and to insert them into the curriculum in a more meaningful manner. From a practical point of view, this situation means that we can estimate what kind of problems can be dealt with successfully by a child who has reached a certain level of cognitive development. To evaluate the relative impact of each of the above variables, well-controlled studies have been carried out. The following section provides one such example.

Nonsemantic Factors of Additive Word Problems

De Corte and Verschaffel (1987b, 1987c) have reported on some of the nonsemantic factors that can affect the choice of solution strategies used by second graders on problems of addition and subtraction. The two factors they have studied involve the order of presentation of the two given numbers, whether the larger or smaller one is given first, and the order of presentation of the two given sets.

Results for Addition Problems

With regard to addition problems, a distinction was made between strategies in which the child begins with the first of the two given numbers (F strategies) and strategies starting with the second one (S strategies). For problems in which the first given number is the smaller one, S strategies are more efficient. By disregarding the given order of the addends and starting with the larger one, the child reduces to a minimum the number or difficulty of the steps in the solution. The three additive problem types were as follows:

- *Pete has . . . apples; Ann has . . . apples; how many apples do Pete and Ann have altogether?* (Combine 1 problem)
- *Pete had . . . apples; Ann gave Pete . . . more apples; how many apples does Pete have now?* (Change 1 problem)
- *Ann gave Pete . . . more apples; he started with . . . apples; how many does Pete have now?* (Change 1 problem but inversed).

Each of these three problem types was presented in two different ways: once with the smaller number given first and once with the larger given first.

The children reported much more frequently that they had solved a problem with a S strategy when it started with the smaller number than when the larger addend was given first (57% vs. 22%). A possible explanation of this difference is, of course, that it is not efficient to start the solution process with the second given number when the larger number is given first. This finding suggests that children's solution strategies are indeed strongly influenced by the location of the smaller and the larger given number.

With respect to problems starting with the smaller given number, the children seemed to find it easier to use the more efficient S strategies in the context of Combine 1 and inversed Change 1 problems (63% and 61%) than when the problem had a normal Change 1 structure (48%). The likely explanation is that the dynamic nature of the change problems invites children to model the described chronological sequence of events in their solution strategy. Therefore normal Change 1 problems elicit the lowest number of S strategies. Because combine problems have no implied action, altering the order of the two given sets and starting with the larger one seems less problematic. The tendency to solve change problems using a strategy that parallels the sequence of events described in the problem can also explain why inversed Change 1 problems elicit more S strategies than normal Change 1 problems.

For the problems in which the larger number is given first, inversed Change 1 problems elicited a higher percentage of S strategies (34%) than normal Change 1 and Combine 1 problems (17% and 14%). Again, children's tendency to start the solution process of change problems with the number identified with the starting set is likely to be responsible for the higher percentage of S strategies for inversed Change 1 problems starting with the larger number given first; for the other two types there is no reason to start the solution process with the second number.

Results for Subtraction Problems

With respect to subtraction problems, it is especially interesting to determine the influence of the distinct task characteristics on the choice of either a direct subtractive or an indirect additive solution strategy. In a direct subtractive (DS) strategy the answer is found by subtracting the smaller given number from the larger one; in an indirect additive (IA) strategy, the child determines what quantity must be added to the smaller given number in order to obtain the larger one. So, in a DS strategy, the solution process starts with the larger given number; in an IA strategy the smaller number is taken first. The two subtractive problem types were as follows:

- *At first Pete had . . . marbles; now Pete has . . . marbles; how many did Pete win?* (Change 3 problem)
- *Pete has . . . toy cars; Pete and Ann have . . . toy cars altogether; how many toy cars does Ann have?* (Combine 2 problem)

Both problem types were presented in two different ways: once with the smaller number given first and once with the larger given first.

The first and most remarkable finding for the subtraction problems was the children's apparently very strong tendency to use IA strategies. Second, as expected, considerably more DS strategies were observed for problems starting with the larger given number than for problems in which the smaller number was given first. This finding supports the hypothesis that the order of presentation of the two given numbers has an influence on the kind of strategies children use to solve subtraction problems. Third, it was expected that the effects of the order of presentation would interact with the semantic structure of the problem. More specifically, it was assumed that the influence of this task characteristic would be greater for static Combine 2 problems than for dynamic Change 3 problems. The results show that Combine 2 problems starting with the smaller or the larger given number indeed elicited considerably different percentages of DS strategies (18% and 43%). Combine 2 problems beginning with the smaller given number were solved only rarely with DS strategies; the percentage of DS strategies was much higher when the larger number was given first. This implies that the order of presentation of the two given sets plays an important role on children's solution strategies in Combine 2 problems. But for Change 3 problems, starting with the smaller given number or with the larger one, the percentage of DS strategies (16% and 22%) and IA strategies (84% and 78%) remained almost the same: Most children continued to apply IA strategies even when the larger number was given first. These findings confirm the hypothesis that the influence of the order of presentation of the given numbers is not alike for all semantic problem types.

Research on additive word problems illustrates that the mathematical tasks cannot be studied on the basis of their mathematical structure alone. Research on the psychological aspects of the task and its demands, together with an understanding of the child's development, is crucial for the understanding of the difficulties that children face in learning arithmetic.

Conclusions

An Emerging Philosophy

What emerges from all these studies is the fact that early arithmetic involves elaborate conceptual schemes of a highly complex nature. No longer are we simply concerned with how far children can count and how well they have memorized their number facts. We are now more interested in studying the evolution of their arithmetical thinking. The various structures that have been uncovered enable us to discern qualitative differences in their thinking. The research results included in this chapter convey an interesting viewpoint about arithmetic and its acquisition: Arithmetic is no longer confined to the mastery of skills, it is rather perceived in terms of thinking processes. And since it is impossible to think for someone else, each learner must appropriate for himself or herself the various thinking processes. Of course, this does not mean that each child has to reinvent the number scheme or the addition concepts: Didactical situations can be introduced that will provide the child with an opportunity to reconstruct these notions. The philosophy that is being brought out here is very much that of constructivism.

Future Research

The cognitive structures that have been described are of a *local* nature, that is, relative to a specific aspect of number or addition and subtraction (e.g., the different number-word skills, the different modeling procedures in addition). One of the open questions to be answered in future research is how all these local cognitive structures interact with each other to create what might be termed *global* structures, one encompassing all the aspects of the number scheme, another encompassing all the aspects of addition and subtraction. At an even more general level, one can ask how these two global structures interact with each other (e.g., How does a child's knowledge of addition affect his or her conception of number?), and how they might fit into a broader *meta-structure,* that of early arithmetic.

Another avenue worth exploring concerns the different ways that local structures can be defined. For instance, many of the cognitive hierarchies appearing in this chapter were established on the basis of success rates. However, in a given sample the success rate for one task might be superior to that of another task without necessarily involving the same children. This type of hierarchy merely provides an index of the relative difficulty or ease of the tasks at hand. A more interesting hierarchy might be provided by verifying whether for children who succeed in a given task *A*, one finds a high percentage of success in another given task *B*, but not conversely . This individual-subject-oriented hierarchy proves to be rather difficult when several tasks are involved and when the sample is relatively large. Sophisticated tools of statistical analysis are needed.

Future research need not be limited to empirical studies assessing the children's existing knowledge. Research of a more pedagogical character might be undertaken in order to explore the impact of cognitive conflict on the child's construction of early arithmetic. Different types of teaching experiments could be designed to investigate the cognitive effect of various instructional interventions. For instance, different problem situations could confront the child with the need to resolve a cognitive conflict created by contradictions between his or her deductive reasoning on one hand and the information gathered from spatio-temporal transformations on the other.

Pedagogical Implications

The pedagogical implications are numerous. The research on early arithmetic has shown the importance of the various counting procedures. Not only does their introduction provide schoolchildren with a greater variety of mathematical activities, but these different counting procedures can be used in many problem-solving situations dealing with number (e.g., finding how many chips are hidden in a row by counting back from a given chip). Many new activities can now be introduced to compensate for the perennial neglect of the ordinal aspect of number. Other numerical activities can provide opportunities for children to reflect and discover the invariance of number under different figural transformations. More generally, this research can convince teachers that children possess a numerical knowledge far greater than is usually believed.

Although the above counting tasks can expand the children's experiences in kindergarten and Grade 1, research results on addition and subtraction can enrich the program of the first three years of primary school. Teachers familiar with some of this research become aware of a much larger range of addition and subtraction problems, of the existence of different classes of such problems, and of their relative difficulty. Moreover, when they realize the multiplicity of the procedures children can use in solving such problems, teachers will be more interested in how they solve the problems rather than limiting themselves to the correctness of the answer. When teachers discover that children are capable of inventing new procedures to deal with challenging problem situations, they inevitably start focusing on the thinking processes, not merely the end product.

3   LANGUAGE AND MATHEMATICS

Colette Laborde
with the collaboration of
John Conroy, Erik De Corte, Leslie Lee, and David Pimm

Aims of Investigations into the Role of Language
in the Teaching and Learning of Mathematics

Educators are now convinced of the important role language plays in school as far as teaching involving communication about knowledge. In fact, the teaching and the learning of mathematics require various language activities such as reading, listening, writing, and discussing. When learning mathematics in school, students are faced with written formulations by the textbook or by the teacher, with the oral discourse of the teacher in the classroom, and with the talk of their classmates. The content of these statements deals with a wide range of objects, ideas, and activities.

The functions of language in the context of mathematics classrooms are those that have been recognized for a long time in the development of thought: Language serves both as a means of representation and as a means of communication. As a consequence, its role in mathematics education cannot be neglected. Among the language issues inherent in attempting to teach and learn mathematics in school, we (as researchers in mathematics education) are concerned with those issues related to the fact that it is *mathematics* that is being taught and learned. The specific context of instruction and the specific teaching content jointly affect the syntactic, semantic, and pragmatic features of the language used in mathematics teaching, which differs from the language of everyday life not only in its explicit aspects but also in its implicit aspects.

Even if the main structures of the mother tongue are acquired by the child at the level of the elementary school, it appears that mastering the language needs a long time. At the end of compulsory education or later, some students may still encounter difficulties in reading or writing texts in specific contexts such as scientific ones, which require a use of the language remote from the ordinary one. The language through which students have access to the mathematical content may be a bar to both understanding and expression.

In this chapter, we consider some of the language problems students encounter when they attempt to learn mathematics in relation to the conceptual problems involved in the acquisition of mathematical knowledge. This point of view has been stressed many times in the International Group for the Psychology of Mathematics Education (PME). In his introductory lecture to the 1981 meeting, Gerard Vergnaud (1981b) focused on the indissoluble link between the *signified* (the mental representation) and the *signifier* (the external representation). As an external representation system, the language used by individuals is connected to their mental representations, that is, to the ways he or she elaborates and organizes knowledge (Kaput, 1987). This interdependence has two aspects. On the one hand, the way in which an individual understands a text or formulates ideas depends on this individual's knowledge and on conceptions about the content to be read or expressed. On the other, language activities, through the specific problems they cause, can lead the individual to consider the objects and relations underlying the discourse in a different way and can also be of help in problem solving. Language activities are of value for at least two reasons: (a) making explicit the relations underlying discourse creates a need to analyze them with sufficient precision; and (b) the necessity of communicating a message to someone requires an awareness that what is obvious or known to the speaker is neither automatically clear nor necessarily known to the listener or reader.

Within the debate concerning the relationship between thought and language, many psycholinguists have shown that formulating in written or oral language cannot be done without abstracting from the context of the objects to be described, without creating sufficient distance from these objects, and without a decentration, so that an interlocutor external to this context could grasp the meaning of the message solely by means of the verbal data. Conversely, this activity of formulation contributes to abstraction, distancing, and decentration: Language becomes a means of regulation and control of both acting and thinking (Beaudichon, 1982; Oléron, 1978; Vygotsky, 1934/1962).

Five main questions are discussed in this chapter:

1. What are the specificities of the language used in mathematics and in mathematics teaching?

2. What problems due to the specificities of language used in mathematics do students encounter when writing or reading mathematics--in particular, in moving from one form of language to another?

3. What is the role of language comprehension in problem solving, or more precisely, how can the wording of a problem statement affect the solution elaborated by the students?

4. What is the role of verbal interaction in mathematics learning?

5. In what way can direct teaching develop students' abilities to read or formulate ideas in mathematics?

The investigations presented in this chapter generally deal with students in compulsory education because the process of language acquisition lasts over a long period of time, mainly concentrated on primary and the beginning of secondary education. Only a portion of the students from compulsory education have access to further education in senior high school and college; language achievment is one of the most important criteria contributing to this selection. These older students do not encounter such important language problems as the younger ones do, although some investigations have shown that certain errors in solving word problems may occur even for engineering undergraduates (see the students-and-professors problem on page 61 below). Although Question 1 requires a linguistic analysis, Questions 2 and 3 require diagnostic research based on psychological and linguistic hypotheses, and Questions 4 and 5 require experiments in classrooms based on hypotheses about the nature of knowledge acquisition in a school setting and about the factors of development of this knowledge. But, in any case, such observations (the results of empirical research) can only acquire a meaning in relation to an explicit theoretical framework.

Theoretical Framework

Our concern with the relationship between language problems and the learning of mathematics requires a theoretical framework that takes into account the fundamental elements of the learning situation in school: the content to be taught, the student viewed from a cognitive and social point of view, and the teacher.

From a stimulating debate within our group in the 1987 PME meeting (Kilpatrick, Sinclair, Vergnaud, D. Wheeler), a general agreement arose about a constructivist approach in a broader sense (see, in particular, the chapter by Vergnaud in this book). "Knowledge is actively constructed by the cognizing subject [in an adaptative and interactive process with her or his environment] . . . that organizes [this] experiential world" (Kilpatrick, 1987, p. 7). Such an approach concerns not only the acquisition of mathematics knowledge but also the development of linguistic abilities themselves. It means that, from a psycholinguistic perspective, the object to be studied is not the discourse itself (e.g., that of the teacher, the textbook, or the students), but the discourse as the result of conceptual activity in a given context and in a given social environment.

In this activity, the speaker involves his or her:

- conceptions of the mathematical objects;
- knowledge of the language in general and of the privileged uses of the language in the given social environment; and

- representations of interlocutors, of the *hic et nunc* situation (moment and place of the linguistic activity), and of the aim of the linguistic activity (giving information, convincing somebody, establishing social relations, etc.). These elements constitute what linguists call the *enunciative* situation.

Generally, the speaker constructs a meaning within the enunciative situation through this conceptual activity. His or her formulations are not transparent, however, even in mathematics--although according to widespread opinion, mathematical discourse is supposed to be especially clear and unambiguous. Understanding these formulations requires by the interlocutor conceptual operations involving conceptions and representations of the various elements of the discursive situation mentioned above. From this perspective, reading, which is often perceived as a passive activity, is considered as entailing the active construction of meaning.

To answer Questions 2 to 5 posed at the outset requires an initial investigation of the language used in mathematics and in mathematics classrooms as well as of the discourse situations in mathematics teaching.

Uses of Language in Mathematics and in Mathematics Classrooms

Halliday (1975) drew attention to the notion of the *mathematics register* as a set of specific uses of natural language in mathematics, including particular lexical and syntactical uses as well as meanings. More precise observations lead us to claim that there are various uses of language in mathematics teaching depending on the enunciative situation. For instance, the way a teacher speaks in the mathematics classroom differs from the language used in textbooks. The teacher's speech contains more implicit statements, more ambiguities that often can be cleared up by means of the teacher's presence, gestures, and possible questions asked by the students. This kind of language has recourse to repetition as every oral discourse does, whereas written mathematical texts avoid redundancy and aim at a concise and compact presentation.

Nevertheless, certain specific uses of language are privileged in mathematics teaching, namely those of the language of the mathematics textbook. The notion of a mathematics register may be referring to these uses. The language of university or high-level textbooks too often serves indeed as norm for both the presentation of schoolbooks and the students' written productions: There is an implicit teaching requirement that compels students to imitate the textbook language and style.

Features and Conventions of the Mathematical Writing System

Written mathematical texts are based on the use of natural language and of a writing system made of signs exterior to the natural language such as $+$, $\times$, and $<$, together with letters or numerals that can be combined according to specific rules in order to create well-formed expressions like $b^2 + c$ or $a = 4$. In most cases these symbolic expressions are embedded in sentences written in natural language:

If $a = 4$ and $b = 6$ then find c such that $b^2 + c = a$.

The symbols used in mathematical writing fall into four main classes: logograms (specially invented signs for whole concepts), pictograms (stylized icons in which the symbol is closely related to the meaning), punctuation symbols, and alphabetic symbols.

Logograms: There are a range of symbols that stand for whole words; examples include @ for *at,* $ for *dollars,* and & for *and.* The digits 0 - 9 are logograms, as are $\approx$, $\sqrt{}$, $\int$, $\div$, and %. Certain letters over the years have become stylized and their literal origins lost, for example, $\int$.

Pictograms: These are mainly geometric icons, a clearly interpretable image of the object itself. For instance, $\square$ for square, $\triangle$ for triangle, $\angle$ for angle, or $\perp$ for perpendicular to. These are referred to as *motivated* symbols.

Punctuation and alphabetic symbols are familiar and hence easy to form and consequently available for use as symbols. To function as a symbol, production must not interfere, and so it must be available as a single item. This is one reason for the conservatism of mathematical symbols.

Below we list surface features that are systematically exploited in order to produce a symbol *system* for mathematical writing; that is, there are certain structural relations that allow symbols to be combined in certain ways to produce new compound symbols. In alphabetic languages, juxtaposition is the sole means by which words are formed from letters, that is, linear concatenation of the elements. Many students see merely a linear string of symbols and are unable to gain access to the systematic aspects of the notation that are widely employed to make important discriminations and carry information.

The following are the main principles that are used to convey information: order of symbols, absolute and relative position, relative size, orientation, and repetition. (Further details and examples can be found in Pimm, 1987.) Students need to acquire access to the conceptual distinctions that are marked by these surface variations. (Examples of difficulties of students when dealing with the structure of algebraic expressions are given by Kieran, this volume.) Skemp (1982) writes: "Though the power of mathematics resides in its ideas, access to this power is largely dependent on its notation" (p. 204).

Syntactical and Lexical Particularities of Natural Language Use in Mathematics

Some syntactic constructions that are rarely used in everyday language are privileged in mathematics. Mathematicians tend to objectify: They wish to avoid any human factor and the subjectivity attached to it. This tendency leads them to use (among other things) the passive mood and complex nominalizations. The double objectives of conciseness and precision imply the use of a sequence of noun complements of noun complements of noun complements, and so on. The tendency also implies the use of long noun clauses that include adjectives that themselves require complements. The use of expressions like "the composition of rotations" (two nominalizations) shows that the actions "to compose" and "to rotate" have given birth to mathematical objects. Simultaneously, both the action itself and the author of the action have disappeared.

"The foot of the perpendicular line drawn from A to D" is an example of a complex noun clause used to express the complex relations between mathematical objects in a concise fashion. One can see how the complex net of relations between mathematical objects requires a subtle use of surface marks of dependence between the elements of the discourse in order to be expressed.

Finally, in a written mathematical text, every element of the discursive chain can be predominantly determined by the other elements of the discourse. Elements of an external situation cannot be taken into account; that is contrary to what happens during oral presentations in the classroom. All these particularities--the density and the complexity of the natural language as well as the use of the mathematical writing system--do not correspond to the usual linguistic habits of primary school or middle school students.

Students' Abilities to Move from One Form of Language to Another

The relationship between natural language and the mathematical writing system in arithmetic or algebra has been a major focus of interest. Students do not move spontaneously from one form to another, as has been observed by several researchers.

Natural Language and Notational Forms in Arithmetic

Conroy (1981) contrasted the oral language forms of the four operations with notational forms for students in the infant school. He inquired whether there was a hierarchy in the various language forms for the four operations. He concluded that there are indeed hierarchies that reflect the language of the classroom but that there are also some language structures peculiar to mathematics that the child has to learn, as these do not occur frequently in the child's natural language. In particular, aural decoding proved to be more difficult than the decoding of symbols.

Not surprisingly, oral instructions to add using the terms "greater than" or "more than" and oral subtraction instructions using "greater than," "more than," and "less than" all proved to be the most difficult language forms for students in their second and third year of primary schooling. Students in their third year found oral multiplication instructions using the terms "lots of" and "times" the most difficult, whereas an instruction like "two fives" was much the easiest. For division, these same students found oral instructions easier than symbolic expressions--probably because they were in the early stages of learning this operation and were not yet familiar with the division notation.

A subsequent study by Relich and Conroy (still in progress) is pursuing how students in the initial stages of learning the multiplication process interpret the multiplication sign and represent multiplication statements pictorially, concretely, and orally. Students' choice of words when reading the x sign was somewhat varied and tended to reflect the preferred words used by their teacher. On subsequent tasks, there is evidence to show how students are grappling with the concept and its representation. For example, 6 × 5 may be represented pictorially as an array, as six groups of five randomly scattered, as 30 items arranged in a row, or in some cases as one group of six and one group of five, with a consequent error in process.

Asked to tell a story about 6 × 5, it is not uncommon for the child to say, "I had one group of five, then another group of five, then a group of ten, and then a group of ten." Intuitive commutativity will also appear, and 6 × 5 may be represented variously as six groups of five or five groups of six. Yet again, 6 × 5 can be represented as six groups of five pictorially but shown as three groups of ten concretely.

There is much to suggest that students' ability to interpret the multiplication sign reflects the nature of their classroom experiences, but the inconsistencies that appear within and across examples suggest that teachers need to be sensitive to the struggles students are having in trying to bring order into their interpretations and representations of symbolic expressions. Establishing a consistency between different representations and expressions seems to be one of the important problems encountered by students. One implication for teaching could be that more attention should be paid to the links between different ways of representing and formulating the same mathematical content.

A study carried out by Putnam, Lesgold, Resnick, and Sterrett (1987) with older students examined the students' understanding of sign-change rules in numerical expressions containing parentheses, together with plus and minus signs before as well as after instruction in algebra (Grades 5, 7, & 9). The study focused specifically on the students' (a) ability to judge the structural equivalence of stories (formulated in natural language) involving adding and subtracting money or combining and changing sets of discrete objects; (b) knowledge of the sign change rules applied to formal expressions like $16 - (8 + 3)$ or $11 - (5 - 2)$; and (c) ability to map the formal expressions and the situations presented in a story setting.

The first result of the study was that the students' strategies of judging the equivalence were completely different for story settings than for formal expressions. From the solutions and the students' explanations, it was shown that the students did not apply their informal knowledge of the quantitative relationships involved in the formal expressions to judge the equivalence of the expressions. They only had recourse to computation-based justifications, surface-level comparisons, or the application (often incorrect) of rules operating on the symbols. The second result was that the students had great difficulties in linking the formal expressions to the situational referents in the stories. Finally, the third result, which may have interesting implications for teaching, concerned the potential power of thinking of the formal expressions in terms of situational referents. Mapping the symbols in the expressions to quantities in the situations seemed to facilitate the application of formal rules, giving a meaning to these rules.

From Arithmetic to Algebra: The Influence of Natural Language

Laborde (1982) and Lee (1987; Lee & Wheeler, 1986) used arithmetical problems asking students to justify and generalize statements about numbers. All these problems had the same characteristics: No algebraic code was proposed in the statement of the problem, and the decision to denote a general number by a letter and to express general relations between numbers by means of algebraic statements using these letters depended on the students' initiative. Lee administered paper-and-pencil tests to 118 Grade 10 students in three Montreal high schools and interviewed some of them working on similar tasks, whereas Laborde used only paper-and-pencil tests with three classes of Grade 9 students in Grenoble.

Example 1:
> A girl multiplies a number by 5 and then adds 12. She then subtracts her starting number and divides the result by 4. She notices that the answer she gets is more than the number she started with. She says, "I think that would happen whatever number I started with." Is she right? Explain carefully why *your* answer is right. [Problem used by Lee.]

Example 2:
> Choose any number between 1 and 10. Add it to 10 and write down the answer. Take the first number away from 10 and write down the answer. Add your two answers.
> 1. What result do you get?
> 2. Will the result be the same for all starting numbers?
> 3. Explain why your answer is right.

[Problem taken from Bell (1976a) and used by Lee and (in French) by Laborde.]

Some common conclusions have been drawn from both pieces of work. First, for most students, numerical instances of generalization carry more conviction than an algebraic demonstration of the generalization. And second, students who can competently handle the forms and procedures of algebra rarely turn spontaneously to algebra to solve a problem, even when other methods are more lengthy and less sure.

During the interviews, Lee also observed that some students were able to translate a relation between numbers given in natural language into an algebraic statement, but without any meaning for it. The conclusion that the girl was right came from successive numerical examples but not from the algebraic statement.

Laborde's analysis of the students' justifications given in natural language seemed to show that the students did not describe static relations between numbers but actions on numbers that are made by themselves or by any person. Their formulations often refer to a time scale defined by the succession of actions they take, as in the following explanation (for Example 2) by a 14-year-old French student who had already encountered algebra:

> *Ma réponse est correcte car si on choisit un nombre quelconque inférieur à 10, et qu'on l'ajoute à 10, et ensuite on retranche ce même nombre quelconque à 10 et qu'on ajoute les 2 résultats, cela nous donnera obligatoirement le nombre 20.*
> [My answer is right because if you choose any number less than 10 and add

it to 10, and if you then subtract this same number from 10 and add both results, you will necessarily get the number 20.]

In the case of this student, it is interesting to notice that his explanation of the reason why his answer is right and his answer to the first question are formulated in the same way: He explains what he is doing to get the result. The only changes he made in the formulation is the move from *je* ("I") in the first question to *on* ("one or you") in the last question and the addition of "necessarily." The general explanation is again the description of a procedure, and the generalization is expressed by "one or you" and "necessarily."

On the same problem, only 9 of the 118 students tested by Lee tried to give an algebraic explanation. The explanations obtained were also expressed in natural language and repeated the procedure given in the problem statement, as in the case of the French students.

It could be assumed that the resistance to having recourse to formulae or algebraic statements would be due to the difficulty the students had in going beyond the context and deleting time and action references. The changeover from natural language to mathematical symbolism eliminates all these elements, which create the actual meaning for the students. The history of the development of algebraic symbolism is made up of steps toward the elimination of meaning (see Kieran, this volume). This interpretation is close to the conclusion of Putnam et al. (1987), who stress the importance of situational referents for the students. In this case, the difficulty in using the symbolic writing system originates from the mental representations of the mathematical objects elaborated by the students.

Other researchers have proposed linguistic reasons to explain the errors students make in moving from natural language to algebraic symbolism, as far as these reasons can be separated from cognitive factors. Linguistic features of natural language can affect the translation of a situation from natural language into an algebraic statement, as in the well-known "students-and-professors problem" (Clement, 1982; Cooper, 1984). The equation $6S = P$ is written by many calculus graduates when asked to produce an algebraic expression for the statement that there are six times as many students as professors. It should be pointed out that students interviewed by Clement indicated that they were aware of the greater number of students than of professors in the problem statement. According to Clement, this *variable-reversal* error appeared to stem from using a left-to-right translation of the problem statement and thus was induced by the linguistic structure of the statement in English.

It can be generally concluded that in agreement with our theoretical framework, cognitive and linguistic aspects intervene simultaneously in the comprehension and in the use of different means of expression: The semantics of an expression are constructed by the student by means of his or her mental representations and of linguistic features of the expression. The role that natural language plays in these processes appears to be very strong. Depending on the context, it can help to give a meaning to formal rules (Putnam et al., 1987) as well as being an obstacle to the writing of formal expressions. It means that these processes, according to which an individual constructs a meaning or uses means of oral or written expression, have to be studied deeply and can be dependent on the context and on the mathematical content underlying the formulations. The formal models of description of linguistic systems are not sufficient to explain these processes. Kirshner (1987) proposes, for example, that the user of algebra does not treat operations as binary, contrary to the formal description of the syntax of algebra.

Problems in Language Comprehension

The role of language comprehension is discussed below in two contexts: problem solving and reading mathematical texts.

Effect of Problem Wording on Solution

Understanding what is to be solved requires understanding the problem statement given in an oral or written form. Task variables in problem solving have mainly been studied for word problems in arithmetic. Among these variables, the wording of the problems appears to influence the students' problem representations and therefore their strategies of solution. The special character of word problem texts leads to an interpretation that differs from the interpretation of the same items in a narrative sequence or in a discourse (Nesher & Katriel, 1977). Some main variables whose role has been made clear by several pieces of research include the following:

- how relations between the given and the unknown quantities are expressed, and in particular the degree to which they are made explicit (Bachor, 1987; De Corte, Verschaffel, & De Win, 1985);
- the order of the items of information (Mestre, 1988);
- the degree of attraction of some expressions or words, such as the priority of numbers over words or the use of keywords like *more, less* related to arithmetical operations, which may be distractors as well as cues (Nesher & Teubal, 1975); and
- the complexity of the syntax and of the vocabulary (Bachor, 1987, Spanos, Rhodes, Dale, & Crandall, 1988).

De Corte et al. (1985) systematically studied the influence of the first variable on students' performance and solution strategy. Consider the following example:

Version 1:
　　Joe won 3 marbles. Now he has 5 marbles. How many marbles did Joe have in the beginning?

In this problem text, it is not explicitly stated that before winning marbles Joe already had marbles. But this problem can be reworded in such a way that its surface structure makes the semantic relations more obvious:

Version 2:
　　Joe had some marbles. He won 3 more marbles. Now he has 5 marbles. How many marbles did Joe have in the beginning?

De Corte and his collaborators systematically tested the hypothesis that rewording simple addition and subtraction problems so that the semantic relations are made more explicit, without affecting their semantic structure, facilitates the solution of these problems by young elementary school students. Two series of six rather difficult word problems were administered collectively near the end of the school year to 89 first and 84 second graders. In Series A, the problems were stated in the usual condensed form; in Series B, they were reformulated as in the above example. The reworded problems of Series B were solved significantly better than the standard problems of Series A.

The following hypothetical explanation of the mechanism that accounts for this finding was proposed by De Corte et al. (1985). The mental representation of the problem constructed in the first stage of a competent solution process is the result of a complex interaction of top-down and bottom-up analysis; that is, the processing of the verbal input as well as the activity of the competent solver's semantic problem schemes (change, combine, compare problems--see Bergeron and Herscovics, this volume) contribute to the construction of the representation. In inexperienced and less able students, these semantic schemes are not yet very well developed, so these students depend more on bottom-up or text-driven processing to construct an appropriate problem representation. Therefore, especially for such students, rewording verbal problems so that the semantic relations are made more explicit without affecting the underlying semantic structure facilitates the construction of a proper problem representation. Moreover, one can argue that these rewordings do not only facilitate the construction of an appropriate problem representation but probably also influence the selection of a correct solution strategy.

It would be sensible to relate the study by De Corte et al. (1985) to the textual analysis theory of *thematic progression*. Linguists have stressed how the progression can affect the understanding of a text: Elements such as the order of information in the phrase and the relations between known and unknown, as well as the transition between known and unknown (Combettes, 1983), belong to the progression. Every element of the phrase contributes more or less to "push the communication forwards" (Firbas, quoted by Combettes, p. 30): The more an element enables the progress of information, the higher is its degree in what Firbas calls "communicative dynamism." Three categories can be distinguished according to this scale:
- *thematic*--elements of low degree, which do not carry new information;
- *rhematic*--elements of high degree, which carry new information; and
- *transition*--elements of intermediate degree.

In a text, a balance has to be found between these three components to ensure a progression that allows for better understanding by the reader. It is clear that a text with a rapid progression (strong rhematic component and poor thematic component and transition) requires a more elaborated construction: The reader has to organize all the new data given by the text and to relate them to what she or he already knows.

Conversely, a text with a more expanded thematic component and transition makes the links between old and new information more explicit and relieves the reader of this organizational work. But a text with too slow a progression (not enough new information and an overemphasis on known information) produces an unnecessary reading burden.

Between the two versions of the problem used by De Corte et al. given above, it is possible to notice how the rapidity of the progression was changed from the first version to the second. In Version 1, the progression is very rapid since each sentence contains two elements high in information ("won" and "3 marbles" for the first, "now" and "5 marbles" for the second) versus both a poor thematic component ("Joe," "he") and a poor transition ("has"). If there is too much high-level information in comparison with the amount of low-level information to enable the reader to structure the set of data, the reader first selects among the high-level information elements. Although "won" is, in fact, an item of very high information level, the student's attention is mainly drawn to the quantitative data "3 marbles" and "5 marbles."

Not only the balance between the three components but also the rhematic component is changed in the modified version. In the first sentence, the rhematic component contains two elements "had" and "some marbles" in increasing order of information level. In the second sentence, the rhematic component is restricted to "3 more marbles." In the last sentence, the rhematic component is constituted of "now" and "5 marbles." From these new text components, the highest information-bearing items that may be retained by students are now "some marbles," "3 more marbles," and "5 marbles." One can notice that the symmetry between "3 marbles" and "5 marbles" of the old text is replaced by a symmetry between "some marbles" and "5 marbles" and by an asymmetry between "3 more marbles" on the one hand and "some marbles" and "5 marbles" on the other. The distinction between state and transformation has been made explicit. This change can explain why the typical addition error (consisting of adding the two numbers 3 and 5) occurred less frequently with the second text. The rewording of the texts toward more explicit information induced a better balance between already known and new information elements, which resulted in the students' taking more items into consideration and in a new way.

The studies by Mestre (1988) on the language comprehension of older (Grade 9) students when solving word problems propose similar features of the wording of the problem that may affect the students' comprehension: the order of the data ("the first important noun in the problem statement serves as a trigger," Mestre, p. 209, or the left-to-right translation process in the students-and-professors problem, mentioned above, p. 61), the misleading use of keywords ("more" is interpreted by the student as a mark of inequality), and implicit problem elements (if the word "equal" is never explicitly used in the problem statement, some students do not interpret the phrase as a statement of equality, Mestre, p. 211).

Reading Mathematics Texts

The texts that students are mostly confronted with are the texts of problems to be solved. The reading of these texts can give rise to difficulties as was pointed out above (pp. 57 and 60). But even if some of the origins of these difficulties are known and even if it is possible to decrease the linguistic complexity of a text, very little is known about the ways in which students read complex mathematical texts and learn from them. Linguistic research on reading assumes that the most widespread reading strategy is what is called the *garden path* strategy (Frazier & Rayner, 1982): The reader begins a text with a hypothesis about its interpretation (this interpretation results from the features of the text, the situation, and the reader's knowledge) and goes on reading while retaining this hypothesis. If the chosen hypothesis enables the reader to construct a coherent local and global interpretation, there is no problem, and the reader is even not aware of the reading hypothesis. In this case, for the reader, each proposition has a coherence in itself, and these propositions are organized globally in a consistent way at the level of macrostructure (Kintsch & van Dijk, 1978). If the chosen hypothesis is inconsistent with the remaining text, that is, if the mobilized knowledge is not compatible with the subsequent information in the text, there is a crisis that makes the reader aware of the hypothesis, leads to a reconsideration of the hypothesis, and leads to a rereading of the text with another interpretation.

Some reading experiments have been carried out with students on longer texts than problem statements; in this case, the scope of the coherence between the elements of the text that the students have to construct is larger, and the garden path strategy is likely to be more visible. In these investigations, the students' reading processes can indeed be described with such a strategy (Guillerault & Laborde, 1986; Laborde, 1987). They also seem to show that students have difficulties in rejecting their first hypothesis and that they prefer to take decisions that are in flagrant contradiction to the text. In this way, information in the text can sometimes be more or less deliberately omitted.

Another important use of reading deals with the activity of extracting information from written texts, whose importance progressively increases as the student gets older. How students use a written text or their textbook as a learning tool or as a help to solve problems has received very little attention. The investigations of Balacheff (1987a) and Laborde (1987) are starting points for further studies on this theme.

Verbal Interaction in Mathematics Learning

Although the oral form of language plays a decisive role in the teacher-student dialogue, little systematic research in the PME group has been carried out about the ways students comprehend the teacher's discourse and the influence that the linguistic features of this discourse can have on the students' understanding. It reflects the practical and theoretical difficulties of research on oral language: The transcription of spoken language needs time to be done very accurately, and the analysis of such transcripts is generally complex because the degree of implicitness of oral discourse is greater than in a written communication and elements of the context constituting the enunciative situation (see p. 56) play a more important role. These constraints prevent the analysis of long pieces of dialogues.

The teacher-student interaction has been studied more than the purely linguistic aspects of this dialogue. From this point of view, Pimm (1987, pp. 50-74) introduced the notion of *teaching gambit* to describe certain dialogue strategies used by teachers where they sacrifice a part of the control over verbal interchanges in the classroom in order to gain other advantages. Some teaching gambits provide long-term advantages such as promoting discussion between students or developing students' judgement and autonomy. But the widely used closed style of questioning in which the teacher calls for a one-word answer from the student has the short-term advantage of allowing the teacher to maintain control of the discourse while giving the student the opportunity to say a central word; it has the disadvantage of allowing a very narrow scope for answers and denying students practice in formulating long explanations.

Verbal interaction *between* students has received more attention within the PME group in recent years. It is worth mentioning that the object of study always concerns purposeful talk (e.g., Uyemura-Stevenson in de Avila, 1988, p. 113; Pirie & Schwarzenberger, 1988) directed to a goal such as solving a joint task of finding a solution to a problem. Such cooperative work in small groups was discussed in the PME working group "Social Psychology of Mathematics Education," whose creation followed the lecture by Bishop (1985) on the social aspects of learning and teaching mathematics. This is not the place to discuss these social aspects (a section of the final chapter is devoted to this topic). Let us only mention that all studies agree on the role that verbalization between students plays in the process of constructing a solution (in addition to the works already cited, see Alston Maher, 1988a; Balacheff & Laborde, 1988; Hoyles, 1985; Morange, 1985; Lambdin Kroll, 1988). Nevertheless, some studies mention that a positive outcome is not guaranteed solely by the verbal interaction between students; the features of the task, the relative status of the students, their *cognitive distance,* and their language proficiency affect the interaction and its effect on the solving process (Laborde, 1988). The study by Pirie and Schwarzenberger grapples with the same questions: Evidence to support the hypothesis that discussion aids understanding has been limited, although this hypothesis is very attractive. Pirie and Schwarzenberger propose a classification of mathematical talk in the classroom for a deeper investigation of the links between discussion and understanding. Pimm also concludes that merely increasing the amount of student discourse in mathematics classrooms may not, by itself, prove beneficial. Discussion could be focused on two aims in the mathematics classroom: helping the teacher to know what the student has in mind and improving the students' oral proficiency. The latter point is discussed in the next section.

Teaching Reading and Language Formulation in Mathematics

Very often students are not explicitly taught reading, writing, or speaking in mathematics because these activities appear natural, developing as a matter of course. Students in mathematics are required to imitate the language of the schoolbook or of the teacher, though they have not been taught how to do this. In fact, the only language formulation situations in which students are placed in mathematics classrooms are aimed at enabling the teacher to evaluate the students' knowledge. The students' formulations are meant for the teacher, who already knows the solution of the problem to be solved and can understand even bad or obscure formulations.

In our theoretical framework (p. 55), we stressed how the features of the discourse situation can affect the nature of the language students use. In particular, the aim of the activity of formulating ideas in language can play a decisive role in the quality of the formulations. Changes made in the aim of the language formulation situations in mathematics classrooms can provide a means of improving student language. Usually in mathematics classrooms, the aim is for the student to convince the teacher that she or he knows how to solve the problem, but the student does not actually give new information to the teacher. Setting up genuine communication situations can be used in mathematics teaching to make the students aware of the necessity of resorting to a language understood by others and having clear, precise, and unambiguous formulations (see Pimm, 1987, for further discussion).

A real communication situation, as defined by Brousseau (1986b), is a situation involving two partners A and B (either A or B can be a group of individuals; B can be a machine). B has to perform a precise task but does not have enough information to do it. A has this information but may not do the task himself or herself. To enable B to undertake the task, A must communicate the necessary and sufficient information in a message (oral or written): The quality of the message is a crucial condition for B to be successful.

In this kind of situation, the function of the language is *message-oriented*, according to Brown (1982). With message-oriented speech, the speaker is goal-directed and wishes to express a particular message to change the listener's state of knowledge. It matters that the listener understands correctly.

Several pieces of research and teaching experiments used these kinds of communication situations to foster the learning of codes or of the language used in mathematics:

- Arithmetical writing (Saada & Brun, 1984; Schubauer-Leoni & Perret-Clermont, 1984).
- Coded technical drawings. *A* produced a drawing that represented a polycubal solid to *B*, who had to build the solid using a set of congruent physical cubes (Bessot & Eberhard, 1986; Gaulin, 1985; Osta, 1988);
- Labelling of geometrical objects (points, straight lines, etc.). *A* sent the receiver *B* a written description of a geometrical figure that had been given

to him or her without any indication of the main elements; *B* had to reconstruct the given figure solely by means of the message (Laborde, 1982). Labelling the main elements of the figure could facilitate the expression of the relations between elements of the figure; it appeared here as an efficient tool of communication

- Geometrical lexicon. Message games like the one described above can also promote the learning of a precise vocabulary in geometry. An example on the learning of the expression "symmetry line" (which is very often for the students only a middle line splitting the figure into "equal parts") is given in Grenier and Laborde (1988).

The explicit teaching of how to read mathematical texts is also rare in the classroom. The ability to read a text with the aim of extracting information should be developed by the school insofar as it will be required in the students' future professional life. Just as in situations of language formulation, reading is performed differently depending on its purpose. Students' reading abilities could be developed in various situations, making them aware of the necessity to analyze precisely what they read, such as (a) giving problems to be solved whose text contains unnecessary data (Semadeni & Puchalska, 1987) and (b) giving texts that present new notions or new words not yet introduced in instruction and asking the students either to solve exercises with the help of the text (Rasolofoniaina, 1984) or to produce a new text presenting these notions that is meant for their peers (Laborde, 1987).

Conclusion

The first concluding remark deals with the diversity of the language aspects in mathematics learning and teaching partially reflected by this chapter. This diversity originates from the complex and indissoluble links between language and both conceptual and social aspects of mathematics learning and teaching that intervene jointly in any language activity as we claimed in our theoretical framework.

The link between students' mental representations and their language formulations have been a major point of concern of several studies. These mental representations about mathematical objects or properties can be an obstacle to moving from one form of expression to another, as in the case of the move from arithmetic to algebra. Several studies stressed the important conceptual work to be done by the students in establishing a consistency between various means of external representations and their frames of reference. This relationship between conceptual aspects and linguistic aspects appears again in the studies that attempted to elucidate how the wording of the problem statements affects the students' solution processes. The construction of a problem interpretation is done by students in interaction with their ability to develop a solution strategy. The student's interpretation when reading a mathematical text is likewise based on hypotheses elaborated by the student about the mathematical content that are dependent on his or her mathematical knowledge. It means that some linguistic features of mathematical text may raise difficulties, especially when the mathematical content underlying the discourse is not familiar to students. Few studies focus on the developmental aspects of the links between language abilities and mental representations. These aspects of the interaction between language and cognition stressed by Vygotsky (1934/1962) should be the concern of further studies.

The teaching of mathematics is faced with the apparent contradiction that language is needed to introduce students to new notions and that language may turn out to be an obstacle to students' understanding. Several studies have recognized the role of talking and discussing among students as a means of developing both language proficiency and mathematical knowledge. Nevertheless, the conditions and the aim of such discussions between students seem to be critical factors of the outcome of this talking activity, though ones not sufficiently investigated. The study of the effect of verbalization and language formulation on learning processes that has been carried out in some of the studies cited needs to be extended, especially the effect on the in-depth functioning of the interaction between language activity and mathematical learning. Some precise features of the communication situation between students have been proposed to allow for the acquisition of new terms or new codes in specific cases. These features involve a social dimension, and it seems that using social characteristics of tasks offers a means of developing language abilities in mathematics, even if the social conditions are not sufficient in themselves to allow this learning and need anyway to be accurately determined. The teaching of formulating or talking mathematically has especially been a point of interest, but very few studies deal with the processes under which students are able to extract information from teaching material on the one hand and the teaching of reading on the other. It is, however, a main objective of education to enable students to cope with written material and to acquire knowledge from it in a society that is undergoing rapid scientific and technological progress.

4 PSYCHOLOGICAL ASPECTS OF LEARNING GEOMETRY

Rina Hershkowitz
with the collaboration of
David Ben-Chaim, Celia Hoyles, Glenda Lappan,
Michael Mitchelmore, & Shlomo Vinner

There are two main "classic" aspects of teaching and learning geometry: viewing geometry as the science of space and viewing it as a logical structure, where geometry is the environment in which the learner can get a feeling for mathematical structure (Freudenthal, 1973). At a more advanced stage, this *geometry environment* acquires a broader sense, without the necessity of a real environment as a basis.

There is a consensus that these two aspects are linked because some levels of geometry as the science of space are needed for learning geometry as a logical structure. This point of view--one that sees the different phases of learning geometry as a developmental process--is intrinsic to most of the theoretical work, research, and instruction that is done in geometry and is the thread that connects the different sections in this chapter.

The various phases of geometry learning raise different kinds of psychological questions. If our concern is geometry as the science of space in general, the initial questions are broad, such as:

- How do children perceive their surroundings?
- What kinds of codes are used in processing visual information?

The questions become narrower if we confine ourselves to visualization; for example:

- What kinds of visual abilities are needed for geometry learning? In particular, how do children create *documentation* of their surroundings and how do they interpret this documentation; that is, how do children describe (verbally or visually) the three-dimensional world, and how do they interpret such a description?

Some of these questions are discussed in the section below on visualization.

Another kind of question deals with the processes of constructing basic concepts (e.g., the main geometrical figures) and the relationships between the elements of a concept and among different concepts. Such questions are discussed in the section on concepts and relationships. Studies within the domain of geometry as a logical structure raise questions about generalization and proof processes; this issue is discussed in the section on conjectures and proofs.

In the last few years there has been a considerable amount of research involving geometry in a computer learning environment. The strong visual elements provided by the computer, its interactive potential, and the way visual objects can be easily manipulated and viewed from different perspectives attract many mathematics educators. Most of the work presented at the meetings of the International Group for the Psychology of Mathematics Education (PME) on this topic has focussed on geometry education and not on the computer for its own sake. There is a common interest in using the computer-student interaction to create learning situations that facilitate the acquisition of visual skills, specific geometrical concepts, or thinking processes. Therefore, we can see research questions and trends as they are reflected in the "mirror" of the computer environment. This reflection raises issues relevant to the questions mentioned above, as well as presenting new insights and new areas of investigation. The computer's special contribution is discussed in each section below.

In order to give some global meaning to the research and instructional work described in this chapter, we preface it with a discussion of some theoretical background.

On Theories and Theory-Based Research

We can distinguish two main approaches in which cognitive research relates to theory. In the first, the *theory-based* or *top-down* approach, the focus of interest is the theory that is supposed to be confirmed or disproved. The geometrical content and tasks that are selected in these top-down investigations are chosen to fit the theoretical model and do not necessarily reflect the common content and processes involved in geometry learning. The second approach, the *bottom-up* approach, takes the content and structure to be learned as the starting point; the understanding and explanation of student difficulties and processes are the main goal. According to this approach, theory is not the basis for research design but is used as tool to explain situations and results raised by the research, when the theory fits (Balacheff, 1987b). Research may still lead to the improvement of theories or even to the formulation of new partial theories.

The above distinction between the two approaches is in a sense an oversimplification, and we can find research work that is "between" the two. However, most of the current cognitive research, including research in geometry by members of the PME group, takes the second approach, whereas the first approach was more prevalent few years ago. Thus the discussion in the following sections of this chapter considers theories and partial theories as research and instructional tools. In the remainder of this section, however, we discuss some features of relevant theories, followed by some examples of theory-based research.

Piaget

In his theory of the child's conception of space (Piaget & Inhelder, 1967) and the child's conception of geometry (Piaget, Inhelder, & Szeminska, 1960), Piaget described the development of the child's representational space. This was defined as the mental image of the real space in which the child is acting, where "mental representation is not merely a recall from a memory bank. It is an active reconstruction of an object at the symbolic level" (J. L. Martin, 1976a, p. 28). This process is not purely perceptual. In Piaget's words:

> Perception is the knowledge of objects resulting from direct contact with them. As against this, representation or imagination involves the evocation of objects in their absence or, when it runs parallel to perception, in their presence. It completes perceptual knowledge by reference to objects not actually perceived. (Piaget & Inhelder, 1967, p. 17)

With some simplification, we can say that Piaget, in his typical way, was interested in the mental transformations from real space to the child's representational space, in those attributes of real objects that are invariant under these transformations and how they changed with age. According to Piaget's theory, the child's early transformations are those that conserve topological attributes of objects (e.g., interior and exterior of a set, boundary of a set, connectedness, and openness and closedness of curves). Only later is the child able to transfer to his or her representational space Euclidean attributes of objects (e.g., length of lines and size of angles). The outcome of these Euclidean transformations is the conservation concepts of length, area, volume, and so forth. It is only at this point that the child, according to Piaget, can succeed with measurement and higher-level tasks.

The volume of research based around Piagetian research has been large and diverse. Some studies (Dodwell, 1959; Lovell, 1959) gave support for his theories, whereas others (J. L. Martin, 1976b; Taloumis, 1975) provided contradictory evidence.

Van Hiele

Whereas Piagetian theory relates mainly to geometry as the science of space, the van Hiele theory combines geometry as the science of space and geometry as a tool with which to demonstrate a mathematical structure. The theory distinguishes sequential levels of geometrical thought (Freudenthal, 1973; Hoffer, 1983; van Hiele & van Hiele-Geldof, 1958; Wirszup, 1976; and many others). In addition, the van Hiele theory suggests phases of instruction that help students progress through the levels. The following is a short description of the levels (Hoffer, 1983; Usiskin, 1982).

- *First level: Recognition or visualization.* The student perceives geometrical concepts in terms of their physical appearance; figures are recognized by their shape as a whole, not by their properties.
- *Second level: Analysis.* The student can analyze properties of figures.

- *Third level: Order.* The student can logically order figures and relationships but does not operate within a mathematical system. Thus, simple deduction can be followed, but a whole proof is not understood.
- *Fourth level: Deduction.* The student understands the significance of deduction and the role of the different elements in the deductive structure. Thus, proofs can be "reinvented" by the student or at least understood.
- *Fifth level: Rigor.* The student can work in a variety of axiomatic systems and is able to make abstract deductions. For example, non-Euclidean geometry can be understood.

Later, the van Hiele theory was modified and reduced to three levels (van Hiele, 1987): the first, the second, and a third level that includes more or less the other three levels discussed above. Other characteristics of the van Hiele theory include the following:

- Memorization is not considered to characterize any level.
- The student advances from one level to the next without skipping any level.
- The levels are discrete and global; that is, the student is at the same level in all contexts.
- Students acting at one level cannot interact with or understand teaching at a higher level.
- The development of the individual's thinking from one level to the next is due to teaching and learning experiences and does not depend much on maturity.

It is clear that van Hiele theory takes the final goal of geometry learning to be the realization of geometry as a deductive structure, with geometry as the science of our environment as a necessary prerequisite.

The van Hiele theory, especially the level model, has attracted mathematics educators and researchers. Most of the van Hiele-based research has been done in the U.S.A. (For Russian work more than 20 years ago, see Wirszup, 1976). The hypotheses that the levels can be identified, are discrete, and form a hierarchy have been researched, and the model's predictive aspects have been investigated. In addition, attempts have been made to use the model as a basis for instruction and textbooks.

The generality and globality of the van Hiele model are both its strength and its weakness. In order to use it in research or instruction there is a need to establish operational tools by which an individual's particular level of thought can be determined. Thus, in most of the van Hiele-based research and instruction, there are efforts to establish such tools (e.g., Usiskin's, 1982, test or Hoffer's, 1981, tables).

Research results showed that in general the levels create the described hierarchy and fit the students' geometrical behavior, with a few exceptions:

- The place of the fifth van Hiele level in the hierarchy is not clear (Usiskin, 1982).

- The discreteness and the globality of the levels are doubtful, meaning that a child can act at different levels in different contexts and can even change level within the same task (Burger & Shaughnessy, 1986; Gutierrez & Jaime, 1987; Mayberry, 1983).

The last finding led to the question of a *broad-based* versus a *narrow-based* curriculum; that is, whether to introduce students to many geometrical concepts and to progress with each of them in parallel up to the third van Hiele level (the usual approach) or to introduce the student to a narrow collection of concepts (e.g., quadrilaterals), progress to the third level, and then deal with other concepts. This question has been discussed intensively at the Geometry Group meetings at PME conferences (Hershkowitz & Vinner, 1987). It deserves research and may have an important role in planning geometry teaching.

A typical use of van Hiele-based research has been to determine the levels of a given population. For example, Mayberry (1983), Matos (1985), and Gutierrez and Jaime (1987), studied the performance of preservice elementary school teachers in their respective countries and found that they usually act at the first or second van Hiele level.

A typical example is provided by the work of De Villiers and Njisane (1987). They conducted a detailed study with high school students. In order to make the van Hiele levels more operational, they developed their own test. The items ranged from simple ones like indicating alternate angles when parallel lines are given or listing the properties of a given figure like a parallelogram to items requiring the interpretation of formal definitions and the construction of formal proofs. Most items dealt with content as commonly found in high school syllabi. De Villiers and Njisane distinguished eight categories of geometrical thought needed to answer the items:

1. Recognition and representation of figure types;
2. Visual recognition of properties;
3. Use and understanding of terminology;
4. Verbal description of properties of a figure or its recognition from a verbal description;
5. One-step deduction;
6. Longer deduction;
7. Hierarchical classification (inclusion relationships); and
8. Reading and interpreting given definitions.

Their analyses led them to claim that Categories 1 and 2 belong to the first van Hiele level, Categories 3 and 4 to the second, and Categories 5 and 6 to the third. They were doubtful with regard to Category 7. According to van Hiele, class inclusion is a relationship between concepts and their attributes and therefore is at the third level. But according to their results, it was the most difficult. (The reduced van Hiele model solves this problem.) As was to be expected, they found that the percentage of students answering correctly at a given level increased with class grade level (maturity, experience, or both?) and decreased as the grade level increased.

Another typical use of the van Hiele model has been in research done in the Logo learning environment. The rationale for this kind of research is based on the following:

- The need to bridge the curriculum gap between the level of elementary school geometry, and the level required for the learning of deductive geometry in high school; here instructional sequences based on the van Hiele model seem to fit very well.
- The fact that Logo can be used as a high-level geometrical learning environment and, therefore, has the potential to be the basis upon which the necessary bridge could be built.

The general question is whether Logo experience can accelerate the child's development through the van Hiele levels. For example, Scally (1986, 1987), using a clinical approach with interviews as pretests and posttests for experimental and control groups, investigated how Logo environment might provide experiences at the second and third van Hiele levels for ninth-grade students starting the Logo course at the first or second level. Her work involved the development of an operational definition of the van Hiele levels for the topic of angle, as a basis for the analysis of student interviews. She analyzed the students' development within each level and between levels and found that the students with Logo experience gained more than the control students.

Ludwig and Kieren (1985) investigated, in addition to the above question, a more symmetrical question, that is, the relationships between geometrical knowledge built according to the van Hiele model and Logo use. They videotaped seventh-grade students' behavior while learning geometrical transformations with Logo. The analysis of the videotapes showed a positive relationship: The experience of using Logo procedures as thinking tools to represent geometrical ideas seemed to facilitate the growth of geometrical ideas from the first van Hiele level to the second.

Visualization and the 3D ↔ 2D Example

Visualization generally refers to the ability to represent, transform, generate, communicate, document, and reflect on visual information. Fischbein (1987) discussed visualization and claimed that "very often intuitive knowledge is identified with visual representation. It is a trivial affirmation that one tends naturally to think in terms of visual images and that what one cannot imagine visually is difficult to realize mentally" (p. 103). He continued by arguing that

visual representations contribute to the organization of information in synoptic representations and thus constitute an important factor of globalization. On the other hand, the concreteness of visual images is an essential factor for creating the feeling of self-evidence and immediacy. A visual image not only organizes the data at hand in meaningful structures but it is also an important factor guiding the analytical development of a solution. (p. 104)

There is a general agreement that visualization is important not only for its own sake but also because the type of mental processes involved are necessary for, and can transfer to, other areas in mathematics (e.g., Bishop, 1989). This general agreement supports the line of thought expressed by Fischbein and is especially relevant, of course, to geometry, in which visual elements form some of the "building blocks." Bishop (1983) distinguished between "the ability for visual processing (VP)" and the "ability for interpreting figural information (IFI)." He described VP as involving "visualization and the translation of abstract relationships and non-figural information into visual terms" (p. 184). If we follow Bishop's distinction, we can roughly classify research on visualization into the investigation of visual processing of the visual domain itself and the investigation of visual processing of nonvisual domains.

In this chapter, we are concerned with visualization in relation to geometry learning, which is in a sense visual processing of the visual domain itself. In this section, we discuss some visual skills that seem crucial in geometry learning. In the next sections, we discuss the role of visualization in processes of concept attainment and in higher-level geometrical processes. For a review of visualization itself and the relationship between visualization and mathematics education in general, see Bishop (1980, 1989).

In attempting to investigate how space is perceived and interpreted by individuals, researchers use a wide range of visual tasks and measures, such as the two-way relation between three-dimensional (3D) objects and their two-dimensional (2D) representations, paper folding, and finding embedded figures. In particular, the 3D ↔ 2D transformation, which is a very necessary skill in geometry learning and its applications, has attracted many researchers (Ben-Chaim, Lappan, & Houang, 1985; Bessot & Eberhard, 1986; Bishop, 1978, 1979; Burton, Cooper, & Leder, 1986; Cooper & Sweller, 1989; Gaulin, 1985; Mitchelmore, 1980a, 1980b, 1983; Mukhopadhyay, 1987). One direction, the drawing of 3D objects, has been investigated extensively and serious difficulties have been reported. For example, Mitchelmore (1980a) defined developmental levels for this ability and used them to classify children's spontaneous drawings of 3D bodies. Most of the subjects were at quite a low level. The research by Ben-Chaim, Lappan, & Houang (1989) is a further typical example of this kind of work. They investigated the ability of adolescents to communicate visual information using the Building Description Task, which consists of a building made up of ten small cubes taped together. The students were asked to describe it to an absent friend. The students' attempts were classified by representation mode (verbal, mixed, graphical). It was found that the students had great difficulty in successfully communicating visual information. The various attempts to describe 3D buildings exemplify problems that children have in representing 3D objects. Figure 1 shows some drawings obtained by Ben-Chaim et al. (1989). Figure 1a supports Mitchelmore's (1983) finding that children have problems in representing parallel and perpendicular lines. Figures 1b and 1c exemplify problems that are related to perception of depth in the drawing of views of the building. It is worth noting that the students' descriptions were fairly equally split between the three modes of representation. Burton et al. (1986) found that in a similar task the majority of preservice teachers produced verbal descriptions.

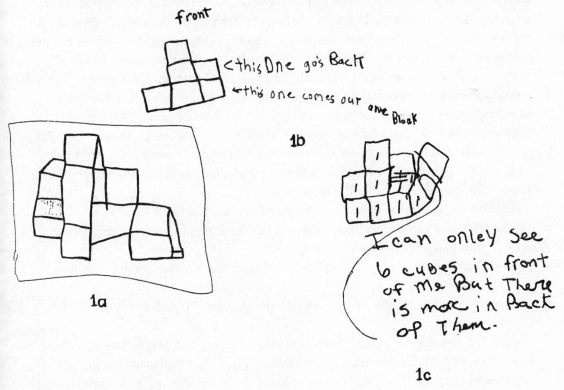

Figure 1. Children's drawings of buildings made of blocks (from Ben-Chaim, Lappan, & Houang, 1989, pp. 132, 137, & 138; reprinted by permission of Kluwer Academic Publishers).

The opposite direction--that is, the interpretation of drawings back to 3D real space--is very important in our modern world, in which we obtain a great part of the information on our 3D environment via 2D media. This direction has been less investigated. The main line of investigations has been explaining how students move from 2D to quasi-perspective drawings, which are assumed to be internalized in much the same way as a view of the actual 3D structures (Metzler & Shepard, 1974). It was found that this direction is also very difficult (e.g., Ben-Chaim, Lappan, & Houang, 1988). Burton et al. (1986) found that in interpretation, preservice teachers prefer visual descriptions of 3D structures to verbal descriptions.

The computer introduces a dynamic dimension into this research on visualization because the representations of 3D and 2D shapes on the screen can be manipulated and transformed in many ways. In addition, the computer enables the researcher to examine microstages of student behavior. For example, Osta (1987) used two commercial programs: MacSpace, in which operations can be performed on the 3D object represented on the screen, and MacPaint, in which operations can be performed only on the 2D *figurative design.* She created instructional sequences of *problem situations* in which the student had to modify 2D transformations performed on the figurative design to make 3D transformations performed on the object represented, and vice versa. The computer puts some constrains that push the student to use geometrical properties of the objects and not just perceptual information. Osta analyzed the solution strategies of Grade 8 and Grade 9 students and found that they tended to be developmental. In the beginning, the students' work was local, dealing with small parts of the whole figure and solving the task by perceptual means only. With experience in such problem solving situations and with the progress they made in the instructional sequences, the students started to consider more global criteria and realized that perceptual means are inefficient and that therefore it is better to use geometrical properties of the 3D bodies.

The 3D ↔ 2D type of research raises questions such as the following:

- What are the factors that influence the description and interpretation of drawings of 3D shapes?
- Can these visualization abilities be acquired or improved by explicit training?
- If yes, what should be included in the curriculum, and how could it be taught?

Concerning the first question, there is much evidence that the factors of culture, experience, and familiarity with the conventions of transforming 3D shapes to their 2D representations and vice versa have considerable effects on the drawing and interpretation of 3D shapes. The three factors are linked: Conventions can be considered as the elements of the "language" formulated by a culture to express and represent the space. Gaining experience is gaining more cultural effect. But different researchers relate differently to these factors. Mukhopadhyay's (1987) study provides an example of a culture and experience effect. She performed an experiment in a *convention free* situation. She asked 8- to 12-year-old children in an isolated village in India who had had almost no schooling and had not been exposed to the conventions of representation common in Western culture to represent solid bodies that were shown to them. She found that their visual representation ability was related to their apprenticeship training in the family occupation (their culture). Thus, potters' children, who worked with 3D solids, produced more complex representations of 3D solids than weavers' or farmers' children (see Figure 2).

There are many other cross-cultural studies that show cultural differences (e.g., Bishop, 1978; Mitchelmore, 1980a). Mitchelmore claimed that these differences "reflect cross-cultural differences in attitude towards the use of spatial models in thinking" (p. 205) and that this attitude is partly revealed by "the degree of geometrical emphasis in the school mathematics curriculum" (p. 205). There is no one conclusion concerning the effects of experience (gained outside or inside school) and of familiarity with conventions. Often, the effect of both is considered to be positive, but there are counterexamples. For example, Mitchelmore (1983) presented evidence that drawing errors persist even when children are very familiar with the usual conventions. Burton et al. (1986) showed that, the same types of difficulties found by Ben-Chaim et al. (1989) persist for adults in spite of increased (albeit unstructured) experience.

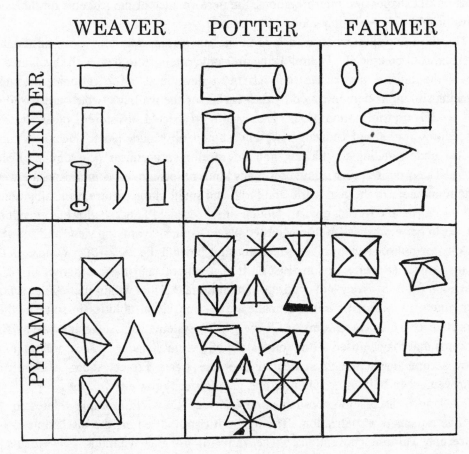

Figure 2. Representations of 3D solids by children apprenticed to different trades (from Mukhopadhyay,   1987).

The research presented raises a second question: To what extent can a direct intervention by instruction improve 3D ↔ 2D transformation ability? The question is important (Gaulin, 1985) but not simple. Bishop (1989) quotes Lean to the effect that these visual "skills (involved in IFI) are trainable given the appropriate experience" (p. 12).

Research work (such as that of Osta, 1987) has given examples of learning experiences through which the child can achieve meaningful development of 3D ↔ 2D transformation ability. Yet there is evidence that shows that the effect of instructional interventions is limited. For example, Ben-Chaim et al. (1988) investigated the effect of direct training. The instructional unit offered concrete experiences with cube buildings and their representation in 2D drawings. The researchers concluded that "fifth through eighth grade students profited considerably from instruction, and the gain was similar for boys and girls despite initial sex differences" (p. 51). Even though the instruction included representations of and familiarity with the drawing conventions for 3D shapes and interpretations, the performance of the students on the same types of items was still only moderately correct.

The third question--what to include in the curriculum and how it can be taught--seems also to be crucial. The following are some attempts to deal with this issue. In the Netherlands a new *visual unit* has been developed. The rationale was that teaching situations demonstrated by the teacher or the textbook, which are based on 3D ↔ 2D transformation skills, are usually restricted to stereotypical drawing procedures (our usual conventions?) and should be broadened to include a more critical understanding of drawing and interpreting procedures (Goddijn & Kindt, 1985). The proposed curriculum uses many kinds of techniques: comparisons between distant and near in the real world and large and small in the picture, vanishing points and the horizon, covering objects, hidden lines, receding lines, shadowing, imagining oneself to be somewhere in the suggested place, front, side and top view, and so forth.

Another holistic-style project, which goes far beyond the 2D ↔ 3D example, is the Agam Project (Razel & Eylon, 1986). It is a general and basic program in visual education for 3- to 7-year-old children. The rationale is that within the regular school curriculum no systematic effort is made to develop visual abilities in spite of their importance. This project aims to fill the gap by systematically teaching basic visual concepts that may be used as the foundation of a visual language. A careful on-going study accompanying the implementation of the Agam Project shows a significant improvement in both visual abilities and geometrical knowledge.

The above discussion of visualization and the 3D ↔ 2D example relates to the positive aspects of visualization. The only criticism that enters the discussion is that concerning cultural conventions. The value of using conventions as elements of language in communicating spatial information is two-sided. On the one hand, we need these "linguistic" elements for communication and for the development of further visual thinking. Therefore there are always efforts to create more linguistic elements to represent the physical world and for visual information processing. The visual language in the Agam Project is an example of such an effort; another example is the *movement notation* created by Eshkol and Wachman (1973), which is the language of the human body movement in the 3D world. On the other hand, in each linguistic development there is some arbitrariness. The developers (individuals or cultures) choose (create) the "building blocks" of the "language," and their choice is not the only possible one by which to construct a language concerning a given ability. The use of a limited number of fixed linguistic elements can therefore put some limitations on the development of that ability. An extreme example in reaction to this "limitation feeling" is the work of artists through the years who have broken the constraints of the accepted visual conventions of their time.

Beyond the visual limitations induced by the use of conventions resulting from the culture, there are visual limitations induced by the individual mind, such as perceptual limitations. These limitations are discussed in the following sections.

Basic Geometrical Concepts and Relations

We include under this heading cognitive aspects of the processes of learning basic geometrical concepts (e.g., angles quadrilaterals, triangles), relationships such as class inclusion, higher-level concepts (the examples of similarity and symmetry), and geometrical measurement.

Basic Concepts

There has been considerable discussion in PME meetings of the distinction between the *concept*--the concept as it follows from its mathematical definition--and the *concept image*--the concept as it is reflected in the individual mind; that is, the product of concept formation processes in the mind (Vinner, 1983). The research aim is to follow the development of the concept image in the individual mind (or in a given population), where the concept provides the frame of reference against which this development is compared and examined. In order to understand better how the student constructs geometrical concept images, and the factors that have an influence on this development, an analysis of the concepts and their mathematical structure is needed. Most of the structure of basic concepts can be considered as conjunction. For example, an isosceles triangle can be seen as a conjunction of the following relevant attributes: (i) a triangle (ii) having two sides (iii) that are equal. (A *triangle* is also a conjunction, but at the stage at which we usually define and learn about isosceles triangles, it is taken as one *chunk*.) The mathematical relationships between the elements of a mathematical concept can be described in the scheme shown in Figure 3.

The concept is derived from its mathematical definition and hence has relevant (critical) attributes (those attributes that an instance must have in order to be a concept example) and noncritical attributes (those attributes that only some of the concept examples possess). The verbal definition itself usually includes a minimal subset of relevant attributes sufficient to define the concept. The definition therefore can be considered as a criterion for classifying instances as positive or negative concept examples. The negative examples (nonexamples) that are relevant to concept formation research and instruction are those that have some but not all of the relevant attributes. Another structural characteristic may be called "the opposing direction inclusion relationship" (Hershkowitz, 1987, p. 240) between sets of examples (concepts by themselves) on the one hand, and their sets of attributes on the other. For example, the set of squares is included in the set of the parallelograms, which is included in the set of quadrilaterals. But if we look at the sets of critical attributes of each of the above sets, we get an inclusion relationship in the opposite direction.

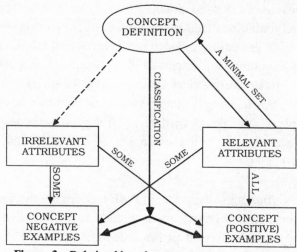

Figure 3. Relationships between concept elements.

In addition to the above structural characteristics, which are not exclusive to geometrical concepts, the examples, the relevant negative examples, and the attributes of the basic concepts in geometry are visual entities. This characteristic gives a flavor of concreteness, which gives them an advantage as a subject for psychological research on concept formation (see the review on mathematical concept formation in general by Sowder, 1980). This kind of research usually investigates a chain or chains in the above structure, which are common to concept formation in general.

Our concern here is the *cognitive processes* that characterize the construction of the basic geometrical concepts themselves. It is clear that in spite of the fact that they can be easily defined structurally in terms of attributes, examples and nonexamples, and so forth, as we did above, these terms are not sufficient to describe the cognitive development of concept images in the mind. Let us describe some main features of this development that have been suggested by research results.

The prototype phenomenon. Vinner and Hershkowitz (1983) and Hershkowitz, Vinner, and Bruckheimer (1987) investigated students' (in Grades 5-8) and teachers' concept images of basic geometrical concepts. The concepts and tasks were sampled from the elementary school syllabus. The researchers found that each concept has one or more prototypical examples that are attained first and therefore exist in the concept image of most subjects. The prototype examples were usually the subset of examples that had the "longest" list of attributes--all the critical attributes of the concept and those specific (noncritical) attributes that had strong visual characteristics: for example, the upright position of a right triangle, the equal sides and angles of square as an example of a quadrilateral, the interior altitude in a triangle, and the interior diagonal in a polygon. This finding is in agreement with other studies (see in particular Rosch

& Mervis, 1975, who investigated the prototype phenomenon intensively in a nongeometrical context). In addition, Vinner and Hershkowitz found that even in *instantaneously formed* concepts, where an *invented concept* was given by a verbal definition without presenting even one example, subjects (students and teachers) produced mostly the same prototypical examples.

The prototype is the basis for prototypical judgement. For each concept, individuals use the prototypical example as a model in their judgement of other instances. Fischbein (1987) calls this kind of judgement "the paradigmatic nature of intuitive judgement" (p. 143). The study by Wilson (1986) illustrates such paradigmatic judgement. She investigated "the relationships between children's definitions of rectangles and their choice of examples" (p. 158), by asking the subjects to define the concept, to identify additional concept examples, and to react to some statements concerning the concept. She found that students wrote definitions that they did not apply when choosing examples or in their reaction to statements. The students' choice of examples was based more on their own prototypes and less on their own definitions.

There are two types of prototypical judgement (Vinner & Hershkowitz, 1983):

- *Type 1.* The prototypical example is used as the frame of reference and visual judgement is applied to other instances (first van Hiele level). For example, in the construction of an altitude in a given triangle, subjects fail to draw examples of altitude that contradict their prototypical concept image of an internal altitude and draw some internal segments that are not altitudes.

- *Type 2.* The prototypical example is used as the frame of reference, but the subject bases his or her judgement on the prototype's self-attributes and tries to impose them on other concept examples. When this does not work, the subject does not accept the figure as an example of the concept. For example, one finds the reasoning: "All the figures, except the square, are not quadrilaterals because they may have equal sides, but they do not have equal angles." This, in a sense, is analytical (second van Hiele level) but erroneous behavior.

The prototype phenomenon and prototypical judgement seem to be mostly a product of visual processes. The prototype's irrelevant attributes usually have strong visual characteristics, and therefore they are attained first and then act as distracters. Another type of judgement is discussed in the following paragraph.

Analytical features. There is evidence that geometrical concept attainment is, at least partially, a result of logical-analytical features. Some examples are the following:

- *Type 3 judgement.* In addition to the two types of prototypical judgement mentioned above, correct analytical judgement is also common (Hershkowitz & Vinner, 1983). This type of reasoning is based on the concept's critical attributes. For example, Figure 4 "is not a quadrilateral because it is not closed, therefore it is not a polygon, and every quadrilateral is a polygon." The frequency of this type of reasoning, which also shows an understanding of the class inclusion structure, is very low in Grade 5 but increases dramatically from Grade 5 to 7. At the same time, the frequency of visual judgement (Type 1) is low but does not disappear completely even among teachers, and the frequency of the prototypical judgement (Type 2) decreases and disappears completely for teachers.

Figure 4. Stimulus figure for judgement task.

- The number of relevant attributes in the conjunctive structure of the concept has a significant effect on task performance (Hershkowitz, 1989).
- Children, at least from Grade 5 on, can construct quite rich and correct concept images by analytical strategies. For example, when the concept definition is given verbally and subjects are requested to use it in order to identify or to construct concept examples, or when the new concept is formed through a sequence of positive and negative concept examples in which trial-and-error learning is modified by immediate feedback and leads to the testing of conjectures and hence to the discovery of the critical attributes (Hershkowitz et al., 1987; Wilson, 1986).

There is some evidence that the construction of the concept image is a mixture of visual and analytical processes. For example, subjects' behavior changed from one concept to the other: Students and teachers who showed analytical behavior (Type 3) in a quadrilateral task failed to identify nonprototypical right triangles.

There are further characteristics of the construction of basic concepts, such as the following:

- A hierarchical order in the attainment of concept examples (starting with prototypical examples and continuing to others, by either visual processes or analytical processes or both) common to the whole population and progressing with experience.
- There are different types of misconception patterns within the same population: (a) *misconceptions that endure*, which have the same pattern of overall incidence from one grade level to the next, and for students, preservice, and in-service teachers (e.g., failure to identify nonprototypical right triangles); (b) *misconceptions that decrease* with concept acquisition, as one might expect (e.g., the Type 2 prototypical judgement); and (c) *misconceptions that increase* with concept acquisition, which are developed

with the learning process (e.g., the concept image of the altitude as an interior segment).

Implications for instruction. Children meet basic geometrical concepts either in a structured way through school experiences or in a nonstructured way from their surroundings, parents, games, and so forth. The principal characteristics of teaching strategies in these situations are: (a) lack of completeness, in which only some of the examples and attributes are presented; (b) lack of awareness as well as absence of knowledge of the existence of further elements (Hershkowitz et al., 1987) on the part of the teacher or even the textbook; (c) lack of awareness of student difficulties and misconceptions in constructing concepts; and (d) generalization of concept attributes (definitions) given (if at all) by the teacher or the textbook, with the learner seen as a passive receptor.

How can we improve the instruction of basic geometrical concepts? It is, of course, desirable for students to develop analytical ability and to base their judgement on critical attributes (definitions) and to overcome the incompleteness and misconceptions with respect to geometrical reasoning that come from visual thinking by itself. Analytical strategies that were mentioned above can be stimulators in the construction of analytical thinking, and we should not underestimate the students' analytical abilities. These strategies, in which the critical attributes and the positive and negative examples (students errors can be used to generate relevant negative examples) are used in different vivid ways, are also very useful in in-service teacher education (Hershkowitz et al., 1987). But we should not use these strategies too early, because children in the early stages create their concept images mostly visually.

How can the formation of a visually-limited concept image be prevented in this visual stage? The answers to this question cover a complete range between two extreme points of view. One extreme, as in Russian studies (e.g., Zykova, 1969), tends to put the blame on the limited visual experience we offer students in the teaching materials and methods and assumes that enriching the visual experience will totally prevent visual limitations. The other extreme puts the blame on the limitations of our perception; that is, individuals will "impose" their visual limitations on their concept image, regardless of the richness of the examples they have met, and hence we will always have limited concept images. We suggest that the answer may lie between the two extremes. However, we have to provide a learning environment that is as "rich" as possible.

The following example shows the contribution that a dynamic interaction with the computer might have on overcoming the orientation effect on the concept image. Shelton (1985) used a computer program in which 2- to 6-year-old children produced sequences of random examples of isosceles or right triangles of different shapes and orientations. After the interaction, most of the children were free from the upright position prototypes and generalized their concept image of triangles to include all triangular shapes and orientations. Thus a rich and dynamic learning environment overcame perceptual limitations. Geometrical software like the *Cabri Géomètrie* (Baulac, Bellemain, & Laborde, 1988), in which a given figure is continuously redrawn when the student moves it around one of its components has a great potential for providing such an environment.

Another interesting question concerning learning and the relationship between the concept and its attributes is raised by Harris (1987). She taught children about geometrical figures using attributes from everyday life. The emphasis then moved from the regular attributes to more "useful" attributes (e.g., the most useful attribute of rectangles in cardboard box manufacturing is that they tessellate). How can this progression be represented in concept formation processes?

Logo and Basic Geometrical Concepts

A very complex question is, What is the effect of programming in a "geometrical language" like Logo on concept formation and vice versa? In studying this question, we have to take into account relations such as those of *procedure to figure, figure to procedure,* and *subprocedure to subfigure* (Hillel, 1986). Research has indicated that Logo can be used as a mean to design rich geometrical environments in which children can act and then with appropiate intervention come to understand a range of ideas and processes concerning geometrical concepts in a personally meaningful way (Hoyles & Sutherland, 1989; Noss, 1987). Children need the opportunity to engage in inductive generalizations, make them explicit in the programming code, and then debug them. However, research has indicated that children do have difficulties with the figure-to-procedure relationship; they do not necessarily use geometrical ideas when working in turtle geometry (Hillel & Kieran, 1987; Leron, 1983a). They can become confused between turtle turn and angle without appropiate didactical intervention (Hoyles & Sutherland, 1989; Rouchier, 1981), and frequently use perceptual rather than analytic cues (Kieran, Hillel, & Erlwanger, 1986). For example, Hoyles and Noss (1987b) used a Logo-based parallelogram microworld to investigate how students come to understand "the essence of the parallelogram" through modifications of the formalism of the given program. They identified a mismatch between students' initial intuitions and their formalized definitions and documented ways in which students progressively became aware of, and generalized, the relationships embedded within the parallelogram procedure by using Logo. They also pointed out the complex relationship that exists between the symbolic code (the procedure) and the figure. Even when student behavior demonstrated close relationships, it was not at a conscious level. It seems that in spite of the potential that programming in a geometrical language is assumed to have in developing dynamic and generalized concepts, the student has great difficulties in modifying programming elements and their visual products. More research is needed before conclusions can be drawn.

Higher-Level Concepts

There has been some quite intensive research into higher-level geometrical concepts and relations. We cite three examples.

Axial symmetry. Grenier (1985, 1987) investigated student conceptions of axial symmetry in French middle schools. She identified variables that effect students' concept images (conceptions) and their performance in axial symmetry tasks. She found that the ability to construct the image of a single dot does not enable students to construct the image of the whole figure. Students use different procedures that give correct answers in special cases only. Variables that affect task performance are the symmetry axis orientation, the respective position of the different parts of the geometrical figure and axis (prototype phenomenon), and the students' age.

Measurement. There is a common point of view about the stages that meaningful learning of geometrical measurement must follow:

A. *Conservation* of the quantity (length, area, volume) being measured;

B. *The meaning of measurement unit and unit iteration* (arbitrary units, standard units, correct use of measurement tools); and

C. Formulae for calculating the quantity being measured.

Figueras and Waldegg (1984) used the Concepts in Secondary Mathematics and Science measurement test for diagnostic evaluation of measurement concepts in 11- to 13-year-old students. The intention was to use this evaluation to guide the design of measuring activities. They found that:

- *Conservation of area* is much easier than *volume conservation* and even easier than *length conservation.*
- More than half of the students used units incorrectly.
- The measuring process (e.g., the use of a ruler for length measurement) as an iteration of equal intervals was done mechanically.
- Most children found areas and volumes by counting units in spite of past experience with formulae, even when counting was much more complicated (e.g., in the case of volume where children have difficulty in visualizing the units that are not shown).
- The performance in measurement tasks dropped when the numbers were fractions.

Figueras and Waldegg developed and implemented learning activities and found that most of the above difficulties inhibited correct learning processes. They claimed that "a fixed measuring system is introduced far too early in the curriculum of elementary school, thus creating a barrier to the complete understanding of the unit concept" (p. 99).

Maher and Beattys (1986) examined the performance on problem-solving tasks dealing with the concept of area in a clinical study of 10- to 14-year-old students. The goal of the tasks was to assist the stages of conservation and unit iteration (Stages A and B above). They found that students "used square unit iteration as an underlying scheme for finding the area of regular shape but did not do so for an irregular region" (p. 168). Most of the children did not apply the concept of area to describe the size of a region, and of those who did, some expressed their answer in linear units.

Douady (1986) developed instructional sequences for teaching and learning the area concept and observed their implementation in the classroom. Conservation was stressed by moving, cutting into pieces, and reconstruction, and by using a grid (two surfaces on a grid that include the same number of squares have the same area). She observed few strategies similar to the above (e.g., counting units and linear strategies). In interviews, she observed and even induced conflicts of conceptions. These conflicts resulted in changes of strategy.

As we can see, the above studies deal mostly with Stages A and B above. The common belief is that instruction usually starts in the middle of Stage B, with the standard units for measuring regular shapes, and this may be the reason why students do not seem to understand the concept of measuring.

Similarity. The Middle Grades Mathematics Project developed and implemented a similarity unit accompanied by research (Friedlander, Fitzgerald, & Lappan, 1984). The concept of similarity was chosen because it (a) seems to provide children with concrete mental images of proportional reasoning and (b) is considered to be one of the more basic ideas in understanding the geometry of indirect measurement, scale drawing, scale models, and the nature of growth. The unit was taught, and its effect was investigated by pretests, posttests, and interviews before and after instruction. "Interviews indicated individual strategies and cognitive levels of thinking about similarity as a result of instruction. The most impressive finding was the lack of consistent strategies within individuals" (p. 127). The categories of strategies that were found correspond to the classifications found in other studies of proportional reasoning and area; for example, additive strategies, strategies based on visualization, and strategies based on counting. The strategy changed with the numbers involved in the ratios. A computer-based microworld on the same topic has recently been designed by Hoyles, Sutherland, and Evans (1989). The findings to date are similar as far as the pretest is concerned to those of Friedlander et al. as described above, but on the posttest there was evidence of an appreciation of the need for consistent strategies; that is, a recognition that identical strategies are appropiate within a given class of problem and indeed that the appropiate strategy in this case is multiplication. This different result can be interpreted as a result of the immediate visual feedback in the face of incorrect strategies.

Higher Levels of Geometrical Thinking--Conjectures and Proofs

As in other areas in mathematics (see Dreyfus, this volume), higher levels of geometrical thinking relate to the inductive process of generating generalizations; that is, making conjectures and all aspects of justifying (proving) generalizations.

In the traditional approach to teaching geometry, the process of inductive discovery, formulated as conjectures, was almost neglected. This neglect was a result of the classical teaching of Euclidean Geometry as the typical example of a deductive system, which has come in for a lot of criticism (Balacheff, 1987b; Freudenthal, 1971; Schoenfeld, 1986). In the words of Freudenthal, "The deductive structure of traditional geometry has never been a convincing didactical success. . . . It failed because its deductivity could not be reinvented by the learner but only imposed" (pp. 417-418).

There is an agreement that the failure of students to reinvent proofs (fourth van Hiele level) or even larger chunks of a deductive system has two main causes: (a) The logical system, in the way it is usually taught, "gives only the end-product of the mathematical discovery and fails to bring about in the learner those processes by which mathematical discoveries are made" (Skemp, 1971, p. 13); and (b) the learner does not have the logical maturity to prove or the awareness of the necessity for proofs (Balacheff, 1987b). It is a common belief now that inductive, empirical discoveries in geometry are necessary because (a) they introduce a discovery aspect; (b) by regarding the generalization as a conjecture in itself, the learner feels the necessity to prove what he or she has conjectured to be true; and (c) inductive experiences are the intuitive base upon which the understanding and the generation of a deductive proof can be built. Schoenfeld (1986) expressed this point in a more symmetrical way, claiming that "the foundation on which geometrical performance is based includes both inductive and deductive competencies" (p. 226). Cognitive research tries to investigate the realization of the above beliefs and raises questions such as:

1. "How do students make the transition from the specific to the general in geometry?" (Yerushalmy Chazan, Gordon, & Houde, 1986)
2. "How do students formalize their hypotheses and generalizations?" (Yerushalmy et al., 1986)
3. Do students feel the need to justify the conjectures they have produced?
4. What are the processes and difficulties that students have in proving?
5. What are students' feelings concerning the role of proof and its validity? (Fischbein & Kedem, 1982)
6. "What are the contexts in which mathematical proof can appear as being an efficient or relevant tool to solve problems pupils have recognized as such?" (Balacheff, 1987b)

Conjecturing Processes and Deductive Processes

The microcomputer brings a new dimension to research and instruction in the ability to make conjectures and their didactical value. Software like the Cabri Géomètrie (Baulac et al., 1988) and the *Geometric Supposer* (Schwartz, Yerushalmy, and Education Development Center, 1985) creates a powerful learning environment for inductive discoveries in geometry, which can be formulated by making conjectures. For example, the *Geometric Supposer* enables users to draw geometrical elements, to make measurements on those constructions, and most importantly, to repeat those constructions on random shapes or on shapes of the users' own construction. In this learning environment, students engage "in significant mathematics-building/hypothesis-generating activities" (Yerushalmy et al., 1986, p. 184). Yerushalmy and her colleagues investigated the first three questions above. They claimed that using this environment "requires considerable teacher planning and effort on an on-going basis [and it is] also demanding of students since it requires that they take on a larger portion of the responsibility for learning" (p. 187). They found that by using the Supposer (a) high school students (low ability level) were able to think about a figure (a construction) in dynamic terms and to see it as a representative of a whole class, (b) it was not easy for the students to make conjectures ("to find patterns in their data and to state those patterns in general terms," p. 188), and (c) the students did feel the need to justify their generalizations.

Balacheff's (1987b) study is an example of research and instruction that takes the unified view that conjecturing and proving are necessary stages of one process and tries to define contexts in which these processes can appear relevant to problem solving (Question 6 above). He constructed a learning situation designed as "a didactical process in which students--about 12 years old--will discover, then formulate as a conjecture, and then finally prove that the sum of the measures of the three angles in a triangle is 180°" (Balacheff, 1987b, p. 8). At the beginning, the students measured the angles of many triangles and compared their results. Then each student had to predict the sum of the angles of the same triangle and to comment on any discrepancy between prediction and the measurement result. The next step was the birth of a conjecture, and the last stage, the construction of a proof as the result of a common effort by the students and the teacher.

In another study, Balacheff (1985) investigated students' conjecturing behavior when their hypothesis is confronted with counterexamples. The expectation was that when a student discovers a fact that contradicts the hypothesis, she or he will abandon the hypothesis and look for another one. Balacheff gave 13- to 14-year-old students the following problem: "Give a way of calculating the number of diagonals of a polygon once the number of its vertices is known" (p. 224). The students worked in pairs and were observed. The observer intervened from time to time, especially in order to give counterexamples to incorrect conjectures. The analysis of the protocols showed that, contrary to the expectation above, overcoming a contradiction is a complex process. Student behavior was somewhat similar to that described by Lakatos (1976). The changes made in the conjectures during the solution process were closely related to the changes in the meaning of the concepts involved in the question.

Another aspect of the relation between hypothesis and proof was examined by Schoenfeld (1982). He claimed that proofs, assuming that meaningful learning of the topic takes place, should be used by students in order to examine the validity of their mathematical hypotheses. Unfortunately, that does not happen. Schoenfeld observed college students who participated in his problem-solving course. All of them had studied geometry in secondary school and were competent at writing elementary geometrical proofs. They were given a geometrical construction problem on which they worked in pairs while their work was videotaped. The analysis of the videotapes showed that the students generated hypotheses by purely *empirical* and *intuitive* means. Moreover, their verification was also purely empirical; that is, their use of ruler and compass was guided more by eye or hand than by systematic analysis. Schoenfeld's conclusion was that "students come to believe that proving is a classroom game, an activity of little or no value outside the artificial classroom environment" (p. 173). As a result, students will not engage in proving in contexts that are even slightly different from the common classroom proof context. The reason can be either that students do not accept proving as a valid form of argument or that they reject it altogether. Schoenfeld put the blame on the curriculum. It might be that the construction of teaching-learning situations, as suggested by Balacheff (1987b), is a possible means to improve the above.

. Kramer, Hadas, and Hershkowitz (1986) developed a microworld by means of which students can carry out classical ruler-and-compass constructions deductively. This microworld consists of objects (dots, lines, segments, etc.), operations on these objects (the basic constructions), and laws governing the application of the operations (the deductive geometry laws), and it is self-correcting and self-regulating. Some 10th graders who worked with the software were observed. The observations indicated that corrections occur at the three levels to which the authors paid attention: the analytic level, the syntactic level, and the level of the application of the rules of geometrical constructions. It is important to mention that it is impossible to carry out a construction with this software without first making a careful analysis, and if the analysis is wrong, the student will have no product or an incorrect product. Also, the fact that the legality of every move is watched by the computer program helps to clarify the rules of the game, something which is not at all clear to many students doing geometrical constructions, as Schoenfeld (1982) pointed out.

Proof and Proof Processes

In Euclidean Geometry, *proof* means a formal deductive argument. In cognitive research we are interested in many types of student proofs (justifications). Balacheff (1982) distinguished between proof--any means by which one becomes convinced that a certain statement is true--and mathematical proof (*demonstration* in French). The latter is proof as accepted by mathematicians. Using this terminology, Braconne and Dionne (1987) investigated students' and teachers' understanding of proof and demonstration. In addition, they examined whether there is a correlation between students' understanding of demonstration and their achievement in mathematics or their van Hiele levels. Like Balacheff, Braconne and Dionne distinguished between various modes of proof in geometry, starting with a naive proof ("you can tell from the figure"--first van Hiele level) and ending with a demonstration. They constructed a questionnaire consisting of 12 different solutions to a single geometrical problem taken from a French geometry textbook. Each solution belonged to one of the five proof modes suggested by the authors. Both the students (aged 15) and their teachers were asked to react to and to evaluate the 12 solutions in the questionnaire. It was found that, for teachers, *proof* and *demonstration* were not synonymous. On the other hand, the difference between the terms did not seem to be so clear to the students.

Perhaps the most natural question about the conception of mathematical proof and its status is the following: Does an individual involved with mathematical proof "clearly understand that a formal proof of a mathematical statement confers on it the attribute of a priori, universal validity and thus excludes the need for any further checks?" (Fischbein & Kedem, 1982, p. 128). In order to answer this question, Fischbein and Kedem constructed two questionnaires, one algebraic and one geometrical. The geometrical questionnaire included the following statement: "ABCD is a quadrilateral and *P, Q, R, S* are the midpoints of its sides. One must prove that *PQRS* is a parallelogram." There followed a proof of the statement. The students were asked if they accepted its general validity. After that, the following question was posed: "*V* is a doubter. He thinks that we have to check at least a hundred quadrilaterals in order to be sure that *PQRS* is a parallelogram. What is your opinion? Explain your answer!" The answers of Grade 10 to 12 students were analyzed and three main categories were found:

1. *Consistently formal.* The students in this category understand correctly the nature of mathematical proof.
2. *Consistently empirical.* The students in this category have an empirical approach to mathematical proofs. They believe that additional checks of particular cases add support to the statement that was proved. The proof itself does not guarantee the absolute validity of the statement.
3. *Basically inconsistent.* The students in this category demonstrate inconsistent behavior, accepting the absolute validity of the proof, on the one hand, but not rejecting the need for additional checks, on the other.

It was found that fewer than 10% of the students were consistently formal and about one third were basically inconsistent. Half of the students could not be classified according to the criteria.

The Ontological Status of Geometrical Entities

The ontological status of geometrical entities is almost a philosophical problem, but it finds its way to the geometry curriculum. The question is whether the geometrical entities are part of the physical world and if not, what are they? There are several philosophical answers to this question. The most common are (a) the formalist approach (mathematics is derivation from axioms and no mathematical reality is assumed); (b) the realist approach (mathematical objects are abstract objects that exist in an abstract world); and (c) the constructivist approach (mathematical objects are psychological constructs formed in the human mind). Vinner (1981) examined the views of teachers and preservice teachers by means of a questionnaire. The answers fell into three categories:

1. Geometrical objects are part of the physical world (34%).
2. Geometry is concerned with the conclusions that can be deduced from certain hypotheses. Hence, geometrical objects do not really exist--the formalist approach (15%).
3. Geometrical objects exist in the abstract world. They are abstract objects-- The realist approach (47%).

The teachers in the first category may have difficulties with certain concepts; for instance, density or infinity of the straight line.

Concluding Discussion

It is not easy to write a conclusion to this chapter because of the varied nature of the research areas and topics. Hence, we have inserted concluding discussions in different places in the chapter. In these, we pointed out future implications for research and instruction. However, there are a few general points and some general areas that require additional research or instructional effort in the future.

- Geometry learning begins when children start to "see" and to "know" the physical world around them, and it can continue to very high-level geometrical thinking through inductive processes or within deductive systems. The characteristics and goals of research and its pedagogical implications change with the level of geometrical thinking. Therefore, it is surprising that most of the research that has been done in the last few years has concentrated on 4th to 10th grade (9- to 15-year-olds). There are studies involving preservice and in-service teachers as well but almost no research on early childhood (excluding such examples as the Agam Project). It seems as though researchers try to learn about processes that start very early by investigating their traces at more advanced stages. If we consider that research methodology is progressing in the direction of detailed observation and documentation of microstages in the learner's behavior and that modern technology stimulates and refines this trend, it would seem very natural to start observations as early as possible. There is still a need to invest effort in research on the evolution of geometrical concepts, geometrical thinking, and the development of visual abilities.
- Concerning geometry and the computer, we saw above that most software that serves research and instruction in geometry involves a high-level interaction with the computer. The computer environment allows the manipulation of specific objects on the screen in ways that assist students to see them as representatives of a class of objects or a class of constructions with invariant properties. An example is the *Cabri Géomètrie*, in which the geometrical figure has a status similar to the status of a variable, changing but keeping all its relevant properties. From this basis, students

are more able to generalize and reflect upon geometric properties. In addition, a computer-based microworld can be built naturally upon visual objects on which operations can be performed according to certain rules. If inappropiate strategies are used, the immediate visual feedback can lead to a cognitive conflict and thus can provoke the student to rethink the solution process and the geometrical analysis of the situation. We have seen how different software stimulates new directions in research and instruction. However, it seems that much more can be done by using existing software, including commercial programs (see, for example, the work of Osta, 1987). In addition, we still need software suited to the development of proof processes and strategies.

- Visualization and visual processes have a very complex role in geometrical processes. We discussed some of this complexity in the context of the formation of basic concepts. There is evidence that this complexity continues at higher levels of geometrical thinking; for example, Hoz (1979) has shown that perceptual rigidity, in which perceptual features of problems act as distracters, affects the ability to prove theorems. There are clues (Yerushalmy & Chazan, in press) that a dynamic interaction with a geometrical microworld like the *Supposer* contributes to visual flexibility. More work is needed to understand better the positive and negative contributions of visual processes.

- There is an agreement that language is one of the main features that characterizes the level of geometrical behavior (Bishop, 1978, van Hiele & van Hiele-Geldof, 1958). Unfortunately, there are only a few studies that touch on this point. Again, we think that there is a need to devote more research effort here.

- The individual inconsistency concerning geometrical tasks found in a few research projects was usually a by-product of the main goal of the research. It seems worth focussing on this feature of individual geometrical behavior in future research. In addition, there is evidence that, on the one hand, some geometrical competencies, such as visual ability, have "a highly individual and personal nature" (Bishop, 1989, p. 14). On the other hand, visual distracters act in very much the same way on different individuals and populations. This question of the nature of visual abilities deserves more research.

- The relationship between geometrical concepts and numbers is another by-product of geometrical research. The effect of number type (large numbers, fractions, etc.) on the performance of geometrical tasks (e.g., similarity, measurement) deserves more investigation.

- Problem situations in which geometry and algebra are combined together are very common in instruction units (e.g., analytic geometry). However, we can hardly find research dealing in such problem situations.
- Last but not least is the mutual feedback between instruction and research. The direction from the real world of the classroom to research seems to be very active. On the whole, researchers take the common instructional content and tasks as the starting point for their research. Yet, in spite of the fact that psychological research in geometry is instruction oriented, with much research work based on or resulting in instructional sequences, it seems that the real effect of research in geometry on curriculum, teaching strategies, and teacher education is sporadic and would reward more intensive and comprehensive effort.

5 COGNITIVE PROCESSES INVOLVED IN LEARNING SCHOOL ALGEBRA

Carolyn Kieran
with the collaboration of
George Booker, Eugenio Filloy,
Gérard Vergnaud, and David Wheeler

School algebra usually begins in the eighth or ninth grade when students are 14 to 15 years of age. The teaching of equations, functions, and the manipulation of literal expressions and equations traditionally signals the start of algebra. But algebra can also be said to begin when students are taught to extract pertinent relations from problem situations and to express those relations using algebraic symbolism. Both of these conceptions of algebra carry with them strands of continuity and discontinuity with respect to the prior arithmetical learning of students. However, even the seeming continuities require a shift in interpretation from what was considered appropriate in arithmetic.

This chapter begins with a discussion of what algebra is, how it has developed historically, and what demands it makes on the learner from an epistemological perspective. This discussion, which includes a description of the continuities and discontinuities of algebra with respect to arithmetic, forms the backdrop for the second part of the chapter, which presents research findings from several cognitive studies in the learning and teaching of algebra. It should be noted that much of the research cited in this chapter considers the current algebra curriculum as a given. Because recent advances in technology force us to reconsider this given, a section describing the algebra research that has been carried out in computer environments is also included. The chapter concludes with some suggestions for further research.

Epistemological Considerations
Algebra and How It Has Developed Historically

Some of the cognitive processes involved in the learning of school algebra find their roots in the historical development of algebra as a symbol system. We summarize briefly here the three evolutionary stages through which algebra has passed (see Harper, 1987, for an informative discussion of these stages). The rhetorical stage, which belongs to the period before Diophantus (ca. A.D. 250), was characterized by the use of ordinary language descriptions for solving particular types of problems and lacked the use of symbols or special signs to represent "unknowns." The second stage, syncopated algebra, extended from Diophantus, who introduced the use of abbreviations for *unknown* quantities, to the end of the 16th century. Harper has pointed out that the concern of algebraists during these centuries was exclusively that of discovering the identity of the letter or letters, as opposed to an attempt to express the general. The third stage, symbolic algebra, was initiated by Vieta's use of letters to stand for *given* quantities. At this point it became possible to express *general* solutions and, in fact, to use algebra as a tool for proving rules governing numerical relations. There is some evidence (Harper, 1987) to support the notion that algebra students pass through these same periods in the evolution of their usage of symbolism. This evidence is presented later.

The development of a specialized symbolic language stripped away meaning from the language in which algebraic activity had been previously expressed (D. Wheeler, 1989). Rhetorical and syncopated algebra were both fairly easy to follow and understand. But the step to a symbolic system eliminated the meanings of individual items and even of the operations acting on them. The power of the symbolic language is that it removes many of the distinctions that the vernacular preserves, thus vastly expanding its applicability. The cost is that the symbolic language is semantically extremely weak, introducing the difficulty for the learner that, by suiting all contexts, the language appears to belong to none.

The debate among British mathematicians in the first half of the 19th century about the nature of algebra (Pycior, 1984; D. Wheeler, 1989) draws attention to an important epistemological problem. One side took the position that algebra is *universal arithmetic;* that is, it deals with quantities and the permissible operations on quantities and its rules are dictated by the well-known properties of quantitative arithmetic. This attitude put into question the legitimacy of the algebraic use of negative, irrational, and imaginary numbers since these numbers cannot be interpreted as measures of quantity. The position of those on the other side of the debate was that algebra is a purely symbolic system dealing with essentially arbitrary symbols governed by essentially arbitrary rules.

According to David Wheeler (1989), when we come to consider algebra in school we find that algebra is derived overtly from arithmetic, that it is presented as a *generalization* of arithmetic. But the pedagogy of high school algebra is not consistent, and there are many covert signs that algebra has its own rules, not necessarily deducible from the rules of arithmetic. The consequence of this confusion is that it leaves many students unsure of the grounds that justify particular algebraic transformations. Is such-and-such a transformation legitimate because it is implied by the behavior of quantities or does it depend on some apparently arbitrary rule? The consequences of this confusion are articulated in a number of research studies described below.

Continuities and Discontinuities Between Arithmetic and Algebra

Arithmetic and algebra share many of the same symbols and signs, such as the equal sign and the addition and subtraction signs. Despite the seeming continuity between arithmetic and algebra, the interpretation given to these symbols and signs is different in algebra. For example, in elementary school the equal sign is used more to announce a result than to express a symmetric and transitive relation. In attempting to solve the problem

> Daniel went to visit his grandmother, who gave him $1.50. Then he bought a book costing $3.20. If he has $2.30 left, how much money did he have before visiting his grandmother?

sixth graders will often write $2.30 + 3.20 = 5.50 - 1.50 = 4.00$. The symmetry and the transitivity of the equal sign are violated. The equal sign is read as "it gives," that is, as a left-to-right directional signal. The interpretational shift that must occur in algebra with respect to the equal sign is precisely that of respecting the symmetrical and transitive character of equality (Vergnaud, 1984, 1988).

Another seeming similarity between arithmetic and algebra is the presence of letters. In arithmetic, 12 m can mean 12 meters, that is, 12 times 1 meter. But in algebra, $12m$ can mean 12 times the number of meters. The letter, say m, now carries a double meaning--depending on the context. It could mean 1 meter or some unknown number of meters. In arithmetic, letters are also used as arbitrary labels, for example, $4m$ to mean 4 monkeys.

Students bring with them to the study of algebra the conceptions and competencies of their experience in arithmetic. These conceptions have to be enlarged and even modified to cope with the demands of algebra. However, there are aspects of algebra that do not represent a continuation of the methods and symbols learned in elementary school arithmetic. One of these discontinuities is the introduction of formal representations and methods to solve problems that, up to then, had been handled intuitively.

In arithmetic, children use informal methods that do not extend to the problems encountered in algebra or to the representation of algebraic expressions. These informal methods have been described by Booth (1984) as being

> (a) intuitive, that is, based upon instinctive knowledge: not systematically reflected upon and not checked for consistency within a general framework; (b) primitive, that is, tied closely to early experience in mathematics; (c) context-bound, that is, elicited by the features of the particular problem; (d)

indicative of little or no formal, symbolized method; (e) based largely upon the operations of counting, adding, and combining; and (f) worked almost entirely within the system of whole numbers (and halves). (p. 37)

Thus, beginning algebra students experience difficulty in representing formal mathematical methods because in elementary school they do not make explicit the procedures they use in solving arithmetic problems. Furthermore, the procedures they use are often informal methods that are difficult to symbolize. Insofar as algebra can be considered as "the writing of general statements representing given arithmetical rules and operations" (Booth, 1984, p. 1), students' nonuse or nonrecognition of these rules and structures in arithmetic leads to somewhat predictable performance in algebra. For example, if students do not recognize that the total number of objects in two sets containing 5 and 8 objects respectively can be written as $5 + 8$ (rather than 13), it is highly unlikely that they will recognize that $a + b$ represents the total number of objects in the sets containing a and b objects.

The difficulty that beginning algebra students experience with representing explicitly the operations that they use in solving problems is, however, only part of the story of the discontinuity between arithmetic and algebra. Arithmetic is primarily procedural. Strings of numbers and operations are not dealt with as mathematical objects but as processes for arriving at answers. In algebra, however, written symbolic representations are often considered objects in their own right, not necessarily representing specific procedures for solving concrete problems. Some of the mathematical objects that are met for the first time in algebra are expressions, equations with unknowns, functions and variables, and monomials and polynomials. However, these objects often incorporate the familiar operation signs. The procedure is, in fact, part of the object. Take the case of the expression $a + b$ referred to above. It represents both the procedure of adding a to b and the object $a + b$. That there is often not a clear distinction in algebra between the process and the object has been characterized by R. B. Davis (1975) as the *process-product* dilemma. Collis (1974) has used the phrase *acceptance of lack of closure* to describe the ability to hold unevaluated operations in suspension.

Another discontinuity with arithmetic occurs in the representation of word problems by equations--equations that symbolize the relationships between quantities occurring in word problems. In arithmetic, children think of the operations they use to solve the problem; in algebra, they must represent the problem situation rather than the solving operations. We can take a simple problem as an example: "When 4 is added to 3 times a certain number, the sum is 40. Find the number." In arithmetic, children would subtract 4 and divide by 3. But representing the problem situation algebraically would require setting up something like $3x + 4 = 40$, which involves the operations of multiplication and addition. That is, to set up the equation, students must think precisely the opposite of the way they would solve it using arithmetic. Thus, not only do they have to deal with representing explicitly the operations they use to solve a problem, but also they must learn to represent operations that are the inverses of the ones they use to solve the problem intuitively.

But it is not only setting up an equation that requires a different way of thinking in algebra. Solving the equation can require solving procedures never encountered in elementary school arithmetic. The procedures for equation solving rest on the principle that adding the same number to both sides of the equation, or multiplying both sides by the same number, conserves the equality. It must be remarked, however, that problems which can be represented by equations such as $x + a = b$, $ax = b$, and $ax + b = c$ can be easily solved by arithmetic methods. Filloy and Rojano (1984) have pointed out that a rupture occurs with equations of the form $ax + b = cx + d$. This equation type allows students to solve word problems that they would not, in general, be able to solve by arithmetic methods. At the same time, the procedures required to solve equations of this type involve transformations that are different from those experienced in the past; so different, in fact, that they seem counterintuitive: operating with an unknown quantity (subtracting ax or cx from both sides). No longer are students able to rely on arithmetic methods, such as numerical substitution and operating on the numerical terms only; algebraic methods are, in fact, necessary. The gap that exists between, on the one hand, problems that can be represented by equations with one unknown and that can be solved by arithmetic methods and, on the other hand, problems that are represented by equations with an unknown on each side of the equal sign and that usually must be solved by algebraic methods has been characterized by Filloy and Rojano as a *didactical cut*. They claim that this gap must be bridged if students are to move from an arithmetical mode of functioning to an algebraic one.

Cognitive Studies in the Learning and Teaching of Algebra

Much of the research on the learning and teaching of algebra has focused on students' ability to cope with the continuities and discontinuities between arithmetic and algebra.

Apparent Continuities Between Arithmetic and Algebra

The Equal Sign

The notion among beginning algebra students that the equal sign is a "do-something signal" rather than a symbol of the equivalence between the left and right sides of an equation (Kieran, 1981) is indicated by their initial reluctance to accept statements such as $4 + 3 = 6 + 1$. Thinking that the right side should indicate the answer--that is, $4 + 3 = 7$--allows them to endow with meaning equations such as $2x + 3 = 7$, but not equations such as $2x + 3 = x + 4$. That older algebra students continue to view the equal sign as a separator symbol rather than as a sign for equivalence is seen in their shortcutting of steps in equation solving and in their staggering of "adding the same thing to both sides":

Solve for *x:*

$$2x + 3 = 5 + x$$
$$2x + 3 - 3 = 5 + x$$
$$2x = 5 + x - x - 3$$
$$2x - x = 5 - 3$$
$$x = 2.$$

This use of the equal sign persists well into college, as documented by Mevarech and Yitschak (1983), who also showed that the students they tested had a poor understanding of equivalence and of the meaning of the equal sign, despite their ability to solve successfully different types of linear equations.

A teaching experiment was carried out with 12- and 13-year-olds (Kieran, 1981) that was designed to enlarge students' view of the equal sign. Students were first asked what the equal sign meant to them and then to give an example showing the use of the equal sign. Next, they were asked to construct several arithmetic equalities, initially with one operation on each side, for example, $2 \times 6 = 4 \times 3$ (the same operation) and also $2 \times 6 = 10 + 2$ (different operations). They went on to constructing equalities with two operations on each side and then with multiple operations on each side (e.g., $7 \times 2 + 3 - 2 = 5 \times 2 - 1 + 6$). The students were able to justify these *arithmetic identities* in terms of both sides being equal because they had the same value. The equal sign began to be seen more as a relational symbol than as a do-something signal. The right side, by this time, did not have to contain the answer but rather could be some expression that had the same value as the left side. Extending the sense of the equal sign within the framework of arithmetic equalities was considered essential in the construction of meaning for equations having more than one term on each side of the equal sign. The next step, introducing the notion of an algebraic equation, is discussed in the New Mathematical Objects section of this chapter (p. 103).

Literal Terms

Children's elementary school experience with letters in equations is often restricted to formulas, such as $A = l \times w$, and to measurement relations, such as 10 mm = 1 cm. It is this latter use of letters as measurement labels that often interferes with students' coming to understand the meaning of variable terms in an algebraic equation. In the latter "equation" above, not only are the letters read as labels, but also the equal sign is read as a preposition--"there are 10 millimeters *in/for* 1 centimeter."

That older algebra students continue to use this label interpretation of literal terms--even in situations where it is not appropriate--was shown in a series of studies carried out by Clement and his colleagues (e.g., Clement, 1982; Clement, Lochhead, & Monk, 1981). In these studies, several groups of university engineering students were presented with the "students-and-professors problem":

Write an equation using the variables S and P to represent the following statement: "There are six times as many students as professors at this university." Use S for the number of students and P for the number of professors.

It was found that 37% of the students answered incorrectly; and of these, 68% represented the problem as $6S = P$. A related phenomenon was reported by Mevarech and Yitschak (1983), who found that 38% of the 150 college students they tested answered that, in the equation $3k = m$, k is greater than m. (See Laborde, this volume, for an extended discussion of some of these notions.)

In algebra, the old arithmetic interpretations of literal symbols are extended to include *unknowns* and *givens*--as occurred in the history of the development of algebraic symbolism. That these extensions in meaning do not take place automatically is shown in several studies. The various interpretations that algebra students assign to letters have been systematically studied by Küchemann (1981) in the large-scale Concepts in Secondary Mathematics and Science (CSMS) evaluation of 3000 British students, 13 to 15 years of age. Using a classification originally developed by Collis (1975), Küchemann found that the majority of students (73% of 13-year-olds, 59% of 14-year-olds, 53% of 15-year-olds) treat letters in expressions and equations as objects; few are able to consider letters as specific unknowns, and fewer still as generalized numbers or variables. For example, 89% of the 13-year-olds who were tested were not successful with the following question--a question that, according to Küchemann, required the concept of generalized number in order to be answered correctly:

$$L + M + N = L + P + N \text{ is true--always/never/sometimes.}$$

Harper (1987) has suggested the existence of stages in the understanding of a literal term as variable and has pointed out that students use literal terms far earlier than they are able to conceptualize them as variables--that is, perceive the general in the particular. In his interviews of 144 secondary school students, by means of questions such as, "If you are given the sum and the difference of any two numbers show that you can always find out what the numbers are," he found evidence in the students' responses of the three types of solutions that can be identified in the history of mathematics. He noted a shift in preference from rhetorical approaches to using the letter to express unknowns to using the letter to express givens as the students grew in mathematical experience and maturity. However, the Vietan use of letters as a means of expressing givens was adopted by only a minority of more able students. That very few students succeed in learning to use literal symbols as a tool for proving rules governing numerical relations was also shown in the study by Lee and Wheeler (1987), where the focus was on students' conceptions of generalization and justification.

In a teaching experiment designed specifically to encourage the acquisition of the notion of a letter as generalized number, Booth (1982, 1983) found a strong resistance on the part of students to the assimilation of this aspect of algebra. She suggests that "the attainment of this level of conceptualization is related to the development of 'higher-order' cognitive structures" (Booth, 1984, p. 88). A study by Chevallard and Conne (1984) illustrates an interesting teaching approach designed to assist those students who have reached a certain level of algebraic maturity to express, by means of literal terms, general rules governing numerical relations. Their sample protocol with an eighth grader begins with a series of questions that ask the student to take any three consecutive numbers and then to square the middle one and subtract from it the product of the other two. That, for several sets of numbers, the result always came out to 1 led eventually to an algebraic proof involving the equation

$$x^2 - (x - 1)(x + 1) = 1.$$

Discontinuities Between Arithmetic and Algebra

New Mathematical Objects

The new mathematical objects met by beginning algebra students are literal expressions and linear equations. Teaching experiments have been carried out by various researchers in studies designed to construct meaning for these objects--in particular to help students resolve the process-product dilemma inherent in these objects.

Expressions. One study (Chalouh & Herscovics, 1988) aimed at overcoming students' inability to accept algebraic expressions as "answers to problems." The problems presented to the Grade 6 and 7 students involved rectangular arrays of dots, lines divided into segments, and areas of rectangular plots--all of the problems having one of the dimensions either hidden or expressed as an unknown quantity. An example of one of the questions that was presented to their subjects is the following: "Can you write down the area of this rectangle?" (see Figure 5).

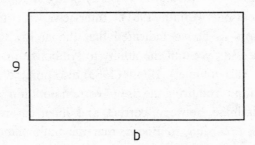

Figure 5. Adapted from rectangular figure presented to subjects in the Chalouh and Herscovics (1988, p. 37) study.

The teaching sequence allowed students to construct meaning for algebraic expressions such as $2x + 5x$. However, the students believed that these expressions were somehow incomplete. They had to express them as part of an equality, such as *Area* $= 2x + 5x$

or as $2x + 5x = $ *something*--suggesting a strong effect from their work with area formulas in elementary school and also pointing out how difficult it is for students to initially make sense of algebraic expressions. Similar findings were presented by Wagner, Rachlin, and Jensen (1984), who found that many algebra students tried to add "= 0" to any expressions they were asked to simplify. The need to transform expressions into equations was also suggested from the results of a study by Kieran (1983), who found that some of the students could not assign any meaning to a in the expression $a + 3$ because the expression lacked an equal sign and right-hand member.

Another difficulty experienced with expressions by beginning algebra students is the correct form of the notation to be used in expressing algebraic answers. For example, one of the items of the CSMS test asked students to determine the area of the rectangle shown in Figure 6.

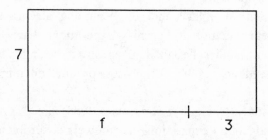

Figure 6. Adapted from the CSMS test item asking students to determine the area of the indicated rectangle (Küchemann, 1981, p. 115).

Forty-two percent of the 13-year-olds responded with $7f3$ or $f21$ or $21f$ or $f + 21$ (Booth, 1984). This item and others from the CSMS test were used in the Strategies and Errors in Secondary Mathematics (SESM) study--a follow-up study that also involved 13- to 16-year-olds (Booth, 1981). Interviews with students who made the same notational errors as above indicated that the ability to verbally describe a method does not necessarily entail the ability to symbolize that method mathematically. (See Laborde, this volume.) Booth (1983) also pointed out that students may respond correctly to items requiring the use of certain notation or conventions and yet be unable to discriminate between correct and incorrect representations. This observation suggests, according to Booth, that the understanding of notation may proceed in stages--a finding echoed by Harper (1987) in his study of students' understanding of the use of literal terms.

Equations. A teaching experiment on the construction of meaning for equations of the form $ax + b = cx + d$ was carried out by Kieran (1981; see also Herscovics & Kieran, 1980). After extending students' notion of the use of the equal sign (see The Equal Sign section of this chapter, p. 100), the next step involved taking one of the student's arithmetic identities (e.g., $7 \times 2 - 3 = 5 \times 2 + 1$) and hiding any one of the numbers. The hiding was done first by a finger, then by a box, and finally by a letter (e.g., $7 \times a - 3 = 5 \times 2 + 1$). Thus, an equation was defined as "an arithmetic identity with a hidden number." Eventually the students hid the same number twice--one occurrence on the left side of the equal sign and the other occurrence on the right-- as, for example, with $2 \times 3 + 7 = 5 \times 3 - 2$ being transformed into $2 \times c + 7 = 5 \times c - 2$. Just as with arithmetic identities, the right side of an algebraic equation did not have to contain the answer but rather could be some expression that had the same value as the left side. For example, equations such as $2x + 3 = 4x + 1$ were described by the students: "If you know what number x is, then 2 times that number plus 3 has the same value as 4 times that number plus 1." This study showed that it is possible to change beginning algebra students' uni-directional and answer-on-the-right-side perception of the equal sign and of equations.

New Procedures

Students are taught several new procedures when they begin their first course in algebra, among them the simplification of algebraic expressions, the solving of equations, and the graphing of functions. Many of the studies that have investigated equation-solving procedures have also had a component dealing with the simplification of algebraic expressions; thus, both are incorporated into the same subsection on equation solving.

Equation solving. Many of the cognitive studies in the learning and teaching of algebra have focused on students' approaches to equation solving. These approaches have been classified into three types: (a) intuitive, (b) trial-and-error substitution, and (c) formal. In general, algebra students are not taught the first two of these types of approaches; students bring them along from their elementary school experience with "missing-addend" sentences (e.g., $2 + \square = 5$).

Intuitive solving approaches include the use of number facts, counting techniques, and cover-up methods. For example, solving $5 + n = 8$ by recalling the addition number fact that 5 plus 3 is 8 would be a use of known number facts. Solving the same equation by counting 5, 6, 7, 8 and noting that three numbers were named after the 5 in order to arrive at 8 would be an example of solving by counting techniques. Booth (1983) has reported the use of both methods among novice algebra students. Bell, O'Brien, and Shiu (1980) have seen students use the cover-up method to solve equations such as $2x + 9 = 5x$: "Since $2x + 9$ totals $5x$, the 9 must be the same as $3x$ because $2x + 3x$ also equals $5x$; so x is 3." Petitto (1979) has remarked that intuitive techniques often do not generalize--as in equations involving negative numbers. She noted that students who used a combination of formal and intuitive processes were more successful than those who used only one of these processes.

The use of trial-and-error substitution as an equation- solving method (e.g., solving $2x + 5 = 13$ by trying different values such as 2, 3, 5, and then 4) is very time-consuming and places a heavy burden on working memory, unless all trials are systematically recorded. As soon as algebra students learn to handle a formal method of equation solving, they tend to drop the use of substitution (Kieran, 1985). Unfortunately, they also seem to drop it as a device for verifying the correctness of their solution (Lewis, 1980). Nevertheless, there is evidence that students who use substitution as an early equation-solving device--and not all of them do--possess a more developed notion of the balance between left and right sides of an equation and of the equivalence role of the equal sign than do students who never use substitution as an equation-solving method (Kieran, 1988).

Formal methods of equation solving include transposing (i.e., "change side, change sign") and performing the same operation on both sides of an equation. Although transposing is considered by many algebra teachers to be a shortened version of the procedure of performing the same operation on both sides, these two solving methods appear to be perceived quite differently by beginning algebra students (Kieran, 1988). The procedure of performing the same operation on both sides of an equation emphasizes the symmetry of an equation; this emphasis is absent in the procedure of transposing.

In a teaching experiment designed to aid students in constructing meaning for the procedure of performing the same operation on both sides of an equation (Kieran, 1983), it was found that students who had begun the study with a preference for the transposing method were not, in general, able to make sense of the procedure being taught, that is, the "same operation" method. The instructional sequence seemed to have its greatest impact on those students who had begun the study with an initial preference for the substitution solving method and who viewed the equation as a balance between left and right sides.

The relationship between formal and intuitive equation-solving procedures was researched by Whitman (1976). She taught 156 seventh-grade students (six intact classes) in one of three ways:

1. Intuitive techniques only: For example, to solve the equation
$$69 - 96/(7 - a) = 37,$$
the student was urged to think:

69 minus what number gives 37?	-- 32
96 divided by what number gives 32?	-- 3
7 minus what number gives 3?	-- 4

Solution: $a = 4$.

2. Formal techniques only: That is, multiply both sides by $(7 - a)$, remove parentheses, combine terms, subtract 259 from both sides, add $69a$ to both sides, and divide both sides by 32.

3. Intuitive techniques followed by formal techniques.

Whitman found that students who learned to solve equations only intuitively performed better than those who learned both ways in close proximity, whereas students who learned to solve equations only formally performed worse than those who learned both techniques. Whitman concluded that formal techniques tend to thwart students' intuitive ability to solve equations.

The effectiveness of concrete models in the teaching of equation-solving procedures was researched by Filloy and Rojano (1984, 1985a, 1985b). Their aim was to help students create meaning for equations of the types $ax + b = cx$ and $ax + b = cx + d$ and for the algebraic operations used in solving these equations. Their principal approach was a geometric one, although they also used the balance model in some of their studies. The geometric approach involved the following kind of statement, accompanied by a drawing of the type illustrated in Figure 7:

> A person has a plot of land of dimensions A by x. Next the person buys an adjacent plot with an area of B square meters. A second person proposes to exchange this plot for another on the same street having the same area overall and the same depth as the first plot of land, but a better shape. How much should the depth measure so that the deal is a fair one?

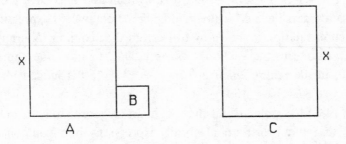

Figure 7. Adapted from the pictorial representation of a geometric situation used to model equations of the form $Ax + B = Cx$ in the Filloy and Rojano (1985a, p. 156) study.

Teaching interviews were carried out with three classes of 12- and 13-year-olds who already knew how to solve equations of the types $x + a = b$ and $ax + b = c$ but who had not yet seen equations of the types $ax + b = cx$ and $ax + b = cx + d$. Filloy and Rojano had hoped to uncover some of the obstacles experienced by students during the period of transition from *arithmetical* equations to *algebraic* ones, that is, during the period of the didactical cut. The interviews revealed that the use of these two concrete models (the balance and the area models) did not significantly increase most students' ability to operate at the symbolic level with equations having two occurrences of the unknown. The well-known equation-solving error of combining constants and coefficients was also seen in this study, in particular with the use of the geometric model. Filloy and Rojano reported that many students tended to fixate on the models and seemed unable to see the links between the operations performed on the model and the corresponding algebraic operations. As a result, the students remained dependent on the model even when it was no longer useful. In fact, students tried to use the model for simple equations that could, more easily, have been solved by the intuitive equation-solving methods that they had used before being taught the new method. They were so fixated on the concretely modeled procedure being taught that they seemed to forget their previously used methods.

Some of the other studies on equation solving have addressed the issue of students' knowledge of the *structure* of equations and of equation solving (Kieran, 1989). Wagner et al. (1984) found that algebra students have difficulty dealing with multiterm expressions as a single unit and do not perceive that the surface structure of, for example, $4(2r + 1) + 7 = 35$ is the same as $4x + 7 = 35$.

Another aspect of structure that beginning algebra students are expected to learn concerns the relationship between operations and their inverses and the equivalent expressions of these relationships. We assume that students entering high school know, for example, that $3 + 4 = 7$ can be expressed as $3 = 7 - 4$ and will be able to generalize this knowledge to equations involving literal terms, thereby becoming aware that $x + 4 = 7$ and $x = 7 - 4$ are equivalent and thus have the same solution. That algebra learners have difficulty in judging the equivalent expressions of the addition/subtraction relation is shown by two errors they commit (Kieran, 1984): In the *switching addends* error, $x + 37 = 150$ is judged to have the same solution as $x = 37 + 150$; in the *redistribution* error, $x + 37 = 150$ is judged to have the same solution as $x + 37 - 10 = 150 + 10$.

Greeno (1982) has pointed out that beginning algebra students are not consistent in the way that they partition algebraic expressions into component parts. For example, they might simplify $4(6x - 3y) + 5x$ as $4(6x - 3y + 5x)$ on one occasion but do something else on another occasion. That a change in the context of a task can lead to a different structuring of the expression was also found by Chaiklin and Lesgold (1984).

Functions. Functions can be viewed as objects or as procedures for computing one magnitude by means of another. The research dealing with functions as objects is treated by Dreyfus (this volume).

Even though students are generally taught about functions later on in traditional courses in algebra, most have already experienced functions at some intuitive level in elementary school. Dreyfus and Eisenberg (1981) investigated these intuitive bases for functional concepts among 440 sixth- to ninth-grade students. They asked questions on image, preimage, growth, extrema, and slope in three representational settings-- graph, diagram, and table of ordered pairs--in both concrete and abstract contexts. It was found that high-ability students preferred the graphical setting for all concepts, whereas low-ability students preferred the tabular setting. Though neither the graphical setting nor the tabular setting specify directly how to compute one magnitude by means of another, the findings of this study suggest that low-ability students may be able to derive this information more easily from tabular settings than from graphical settings.

The difficulties experienced by many algebra students with graphical representations of functions have been illustrated in several research studies (e.g., Clement, 1985; Janvier, 1981; Kerslake, 1977; Ponte, 1984). Kerslake, for example, has shown that in reading graphs representing time and distance, students confused the graph with the shape of the road itself. She also found evidence to suggest that graphs related to real-world situations are no easier for students than graphs that are related merely to symbolic, decontextualized equations. Many students seem unable to make the connection between numerical data and graphs on the Cartesian plane. Similar difficulty has been identified regarding the number line, notably in dealing with scales (Vergnaud & Errecalde, 1980).

The distinction between function as an object and function as a procedure is illuminated by the findings of a study carried out by Sfard (1987). She attempted to find out whether sixty 16- and 18-year-olds who were well-acquainted with the notion of function and with its formal structural definition, conceived of functions operationally or structurally. An *operational* conception, according to Sfard, is one that views a function as an algorithm for computing one magnitude by means of another. A *structural* conception is one that views a function as a correspondence between two sets. The majority of the students who were tested conceived of functions as a process rather than as a static construct. In a second phase of the study involving ninety-six 14- to 17-year-olds, students were asked to translate four simple word problems into equations and also to provide verbal prescriptions (algorithms) for calculating the solutions to similar problems. They succeeded much better with the verbal prescriptions than with the construction of equations. These findings support the results of a previous study (Soloway, Lochhead, & Clement, 1982) that showed that students can cope with translating a word problem into an "equation" when that equation is in the form of a short computer program specifying how to compute the value of one variable based on another. These findings suggest a predominance of operational conceptions among algebra students.

The results of the Sfard study raise some important questions regarding the teaching of mathematics. The symbols and definitions taught in high school are often structural, not operational--in the sense of Sfard. This traditional approach to teaching algebra seems not to be very effective. According to Sfard (1987),

> If an operational conception is indeed the necessary first step in an acquisition of a new mathematical idea, we can probably precipitate the learning by fostering the student's understanding of processes and algorithms *before* translating them into structural definitions; this can be done by incorporating computer programming into mathematics courses. (p. 168)

Sfard's suggestion regarding the integration of computer approaches into the teaching of algebra is reflected in a few recent studies that are discussed below.

Learning of Algebra in Computer Environments

In the preceding sections we have witnessed some of the major obstacles in the transition from arithmetic to algebra and have also seen, in several cases, how despite instruction based on a structural perspective, many students continue to prefer and seem even to understand more clearly procedural (i.e., operational in Sfard's sense) approaches. The advent of the computer onto the educational scene offers the potential not only to build upon these understandings in helping students to bridge the gaps between arithmetic and algebra but also to lead them towards developing a conceptualization of some of the structural features of algebra.

These pedagogical rewards are possible because the computer can be used to create special learning environments that would be difficult, if not impossible, to achieve in algebra classrooms where no computer is available. For example, some computer environments call for programming or at least some sequence of code, thus bringing to the fore procedural aspects of mathematical knowledge and the opportunity to compare what was predicted with what results; other environments focus on the links between various representations of mathematical objects and thereby emphasize structural features; still others combine both.

Research on the learning of algebra in computer environments is still in its infancy. Since we have not yet been able to see what the long-term effects of different computer-supported interventions are, the findings presented in this section should be taken primarily as indications of areas in algebra where the use of computers has already yielded interesting results and where further research is likely to be even more fruitful.

A conceptual area where there have been quite a few studies involving computer approaches is the topic of variable. Samurçay (1985) investigated the concept of variable that develops in 9- to 16-year-olds in the various programming environments of Logo, Pascal, and LSE. She found that the concept of variable does not occur spontaneously in young students working in a Logo environment; it necessitates didactical intervention. The Logo Maths Project (Hoyles, Sutherland, & Evans, 1985) incorporated teacher intervention into its study of how 11- to 14-year-olds develop the concept of variable in a Logo programming context. Sutherland and Hoyles (1986) found that students need to experience variables in many different situations before a synthesis can take place. At the end of the 3-year study, students were interviewed to see whether they had been able to make links between the uses of variable in Logo and in algebra. Some of the interview questions were taken from the CSMS questionnaire mentioned earlier in this chapter. Sutherland (1987a) reported that students' experience with Logo enriched their understanding of variable in an algebraic context.

Similarly beneficial results have been reported by Thomas and Tall (1986), who investigated the learning of the concept of variable among 12-year-olds in a BASIC environment. It is to be noted that all of these studies occurred in programming environments, which suggests that the meaning developed by these beginning algebra students for variables was primarily a procedural one (input/output). However, in a Logo context, the use of variables can imply more than this; when incorporated into

Logo procedures, the use of variables implies an ability to generalize a relationship between quantities and to express that generalization in a formal language. This latter aspect is taken up again in the concluding section.

In contrast to the above studies where the focus has been mainly a procedural one, other studies have explicitly emphasized structural approaches to algebraic concepts. For example, P. W. Thompson and A. G. Thompson (1987) worked with 8 seventh graders over an 8-day period at the end of the school year using a special computer program that involved expressions and equations in two formats: usual symbolic form and expression trees. To change an expression by the use of a field property or other transformation, the students had to first choose the transformation and then place a pointer on top of the operation to be transformed in the *tree representation of the expression.* Thompson and Thompson report the rapid development of these novices' understanding of the structure of expressions and equations in this environment. They also describe how students came to see variables as a replacement not only for numbers but also for subexpressions.

Other studies that have focused on multiple representations include the work done by James Kaput, Judah Schwartz, Michal Yerushalmy, and others at ETC (Educational Technology Center, Cambridge, MA). One of these studies (Yerushalmy, 1988), a pilot involving seven mixed-ability students from seventh to ninth grade, included the use of *Resolver*--a combined symbolic and graphical environment allowing the user to transform algebraic expressions and providing feedback on the correctness of the transformation. Yerushalmy found that, although the graphical feedback was convincing to some students, it was much less effective than she had expected. She pointed out that further research with very carefully planned teacher intervention is needed in this area of computer-supported learning.

Some other studies with a major computer component have focused on developing students' understanding of the identification of points on the number line and in the Cartesian plane. Rogalski (1985) presented 11- to 13-year-olds with a target game that required students to provide one or two numbers corresponding to the point on the number line or point in the plane where the target was positioned. If the numbers suggested by the student were too large, feedback was given on the graphics screen in the form of a trace or message. It was found that students were able to succeed merely by means of approximation strategies and did not have to use proportionality-- something that Rogalski had hoped to see. The interest in this study lies in the fact that the students were able to develop an understanding of scales and intervals, based on the screen feedback. This awareness seems more difficult to acquire in a noncomputer environment (Vergnaud & Errecalde, 1980).

One final study to be discussed in this section is the project of Fey and Heid (Heid, 1988; Heid, Sheets, Matras, & Menasian, 1988). This is a computer-intensive, "functional approach to problem solving" algebra curriculum that has been tested with entire classes of first-year algebra students. It includes the use of many different kinds of software, for example, curve-fitting programs, generators of tables of values, symbolic manipulators, and function graphers. The curriculum centers on

> the use of these computer tools to (a) develop students' understanding of algebra concepts, and their ability to solve problems requiring algebra, before they master symbol manipulation techniques, and (b) make the concept of function a central organizing theme for theory, problem solving, and technique in algebra (Heid et al., 1988, p. 2)

End-of-year interviews with both project and control students showed that the project students outperformed their counterparts on such mathematical modeling goals as constructing, interpreting, and linking representations. They also surpassed the conventional classes in improvement of problem-solving abilities and did as well on a department final examination--a test of ability with traditional algebraic manipulations.

Concluding Remarks

In this chapter, we have provided an epistemological analysis of school algebra in terms of the continuities and discontinuities between arithmetic and algebra and have also presented research findings documenting the difficulties that students encounter in making the transition between these two areas of school mathematics. A conclusion that can be drawn from several of these studies is that algebra students, not only the beginners but the older ones too, seem to find it considerably easier to make sense of teaching approaches that focus on procedural explanations and conceptions (i.e., operational in Sfard's sense) rather than structural representations. Thus, if we consider that competence in algebra requires both, an area where further research is warranted is the interface between procedural and structural conceptions of algebra.

As we have seen, some of the recent studies with a computer component have implicitly begun to explore parts of this area. One example is research that is focusing on the conceptual links between equation, tabular, and graphical representations of relationships. Another example is research that is looking at how experience with variables in Logo procedures can develop the ability to generalize relationships between quantities and to express those generalizations in a formal language. This is a very promising avenue of research for it attempts to straddle procedural and structural conceptions in algebra.

In conclusion, the research on students' ability to cope with the continuities and discontinuities between arithmetic and algebra has shown that many of the errors that students make in algebra are the results of reasonable, though unsuccessful, attempts to use or adapt previously acquired knowledge to a new situation. One of the several questions that remains to be researched is whether meaning for the structure of algebra can be built upon students' apparent proclivity for procedural (operational) approaches. By attempting to resolve this issue, we may find ourselves in a better position to help our students make the transition from arithmetic to algebra.

6 ADVANCED MATHEMATICAL THINKING

Tommy Dreyfus
with the collaboration of
Michele Artigue, Theodore Eisenberg,
David Tall, and David Wheeler

This chapter presents the principal currents and results in recent cognitive research on advanced mathematical thinking processes. The ambiguity in the phrase *advanced mathematical thinking processes* is intentional: The word *advanced* can and should refer to either *mathematics* or *processes* or both. Advanced mathematical thinking processes thus include on the one hand, thinking about topics in advanced mathematics, which in this context means mathematics beyond Euclidean geometry and intermediate algebra, and on the other hand, advanced processes of mathematical thinking such as abstracting, proving, and reasoning under hypothesis.

Although the domain of mathematics subsumed under the term *advanced mathematics* is vast, only a restricted number of topics are commonly taught in upper secondary and beginning college classes. Far more hours of teaching and learning are spent on analysis than on any other topic in mathematics, and it is therefore not surprising that the majority of research on the teaching and learning of advanced mathematics has been concerned with topics in analysis, among them functions, differentiation and integration, and differential equations. This tendency has been strengthened by the fact that the teaching of analysis poses a large number of nontrivial problems.

There do not seem to be clear-cut characteristics that set advanced mathematical concepts apart from those in elementary mathematics. Each advanced concept is, however, based on more elementary concepts and cannot be grasped without a solid and sometimes very specific understanding of these. For example, Confrey (1988) makes the point that the usual conception of multiplication as repeated addition is insufficient for constructing the concept of exponential function. Moreover, some concepts only become meaningful within a structure, for example, a vector or a group element. Thus the concepts of advanced mathematics carry an intrinsic complexity. Students cannot understand what a differential equation means unless they have well understood the concepts (rather than the techniques) of differentiation, nor can they grasp the ideas behind the solution methods without understanding integration, linked to visual and numerical ideas. Similarly, differentiation and integration assume an understanding of the concept of function, and the concept of function assumes an understanding of the notion of variable, which in turn presupposes the number concept. This progression leads to a network of interrelated ideas, each idea integrating some of the more elementary ones into an added structure that in itself may be complex; as an example, a function is not only a variable but two variables that stand in a relationship that must obey certain rules. In order to say what differentiation is, we need to consider a function as an object: Differentiation generates a new function from a given one. Other concepts are similarly built from previous ones. Thus each concept has a high degree of complexity and, on top of that, can only be comprehensively understood in a network together with several other concepts. It is precisely this complexity of the concepts that tends to make it difficult for students to grasp them as entities.

The term *process* may refer to mathematical processes carried out by a person actively involved in doing mathematics. There is no clear-cut characterization that sets apart certain processes in mathematics as being advanced while others are not. Abstraction may be one of the first examples that comes to the mind of someone looking for advanced mathematical processes. Abstraction could tentatively be defined as the replacement of concrete phenomena by concepts whose existence is confined to the human mind. A little reflection soon leads to the insight that abstraction as a mathematical thinking process is by no means confined to advanced mathematics. A child learning about, say, the concept of number is required to carry out processes of abstraction repeatedly. Considering other mathematical processes such as analyzing, categorizing, conjecturing, defining, formalizing, generalizing, proving, or synthesizing leads not much further. All of them may and do occur to varying extent also in elementary classes. They do, however, receive a different weight and become much more frequent in more advanced mathematics. The processes of formalizing, defining, and proving, although often marginal in the lower grades, become of central importance in, say, a course in advanced analysis. The demands made on the student's powers of analysis and synthesis or abstraction increase considerably with the progressive *mathematization* of courses.

Abstraction was considered above as a mathematical process that takes on particular significance in the learning of advanced mathematics. In the learning experience, this process takes on, in addition, quite a different perspective, namely a psychological one. Many processes that are important in the mathematics classroom have both a mathematical and a psychological component. In addition to abstracting, the processes of particular importance in advanced mathematics are representing, forming concepts, inducing, and visualizing. It is one of the main goals of cognitive research in mathematics education to study the psychological aspects of these processes, as they relate to the learning of mathematical concepts. The word *learning* is of particular importance here. In many cases, the way in which a mathematical concept is learned is quite different from the way it is best presented in a formal logical manner. In fact, the logical presentation often contains serious cognitive obstacles to the learner (see Laborde, this volume, for a more extensive treatment of this point). One well-known example is the notion of limit, and some of these obstacles are be presented below.

Cognitive research in advanced mathematical thinking has two kinds of precursors. One kind is presented in papers by mathematicians with an interest in education, among them Lebesgue, Poincaré, Hadamard, Halmos, Hilton, and Thom (see, e.g., Hilton, 1976; Thom, 1972). These first-class mathematicians thought about teaching mathematics, and they were very serious about it. They took the mathematical content and its structure as basis for their thoughts. They did not sufficiently take account of the student involved in learning the mathematics or of the details of his or her understanding and how it is acquired. They did not investigate students' thought processes. In spite of their concern and their efforts, some serious crises have arisen in the teaching of college mathematics; for instance, the present crisis in the teaching of calculus (Steen, 1987). One of the reasons for such crises is precisely that in college mathematics teaching there is usually no consideration of cognitive processes but only of mathematical content.

The second kind of precursor is cognitive research in mathematics learning that dealt with elementary topics in mathematics, mainly with the acquisition of early number concepts. Such research was prevalent in the research community of the International Group for the Psychology of Mathematics Education (PME) until the mid-1980s. One reason for the prevalence may be that cognitive research in mathematics learning is an interdisciplinary undertaking that asks for the collaboration of psychologists and mathematicians; this collaboration necessitates a topic on which they can develop a common language. During the 1980s the mathematics being investigated by cognitive methods have become more and more advanced. A considerable part of PME work is now being done in algebra; moreover, at recent PME conferences, an increasing number of papers have been presented on advanced topics (see the reference list). In addition, a working group on advanced mathematical thinking was started within PME in 1985 and has been active at every conference since then, as well as between conferences. This chapter has been strongly influenced by work done in that group. An in-depth discussion of the differences between advanced and elementary mathematical thinking by the working group has failed to provide an unequivocal answer. Just like abstraction, most aspects of mathematical thinking are, in appropriate forms, already present at very early stages in mathematics learning. But none of the participants in the discussion had any doubt that proving by induction must be classified as a very advanced form of mathematical thinking. One fundamental reason was that proving by induction implies reasoning under hypothesis in a much more essential way than other proof techniques. The working group has concentrated on psychological constructs such as representations and on the description of processes that occur in the learning of advanced mathematical topics. Some of these themes are discussed in the next section. In the following section, we attempt to describe some of the more important work that has been done in the investigation of cognitive processes associated with the learning of specific topics in advanced mathematics. In accordance with what has been said above, the topics of *function* and *calculus* were chosen as representative, and work on these topics is presented in some detail and is shown in the light of the chosen themes. In a later section, studies are described that have investigated difficulties students exhibit when they are asked to start thinking as mathematicians do. Here as well, representative studies on chosen aspects are presented rather than a review of the entire domain. In particular, studies on *proving* and *problem solving* are discussed. In the final section, a number of educational implications are drawn and open research questions formulated.

Themes

Research results of cognitive studies in the learning of advanced mathematics are generally fairly coherent in the sense that the same or analogous results are found among students in different countries and among students at different levels of education and ability. For instance, difficulties with producing and interpreting graphs are all pervasive; student errors when solving limit problems in beginning calculus are the same in France, England, Israel, and Poland; and quantifiers are one of the stumbling blocks in understanding proofs. Several different tentative theoretical frameworks have been proposed and often successfully applied to the interpretation of the same learning difficulties. It would not be doing justice to the research that has been done if the experimental results were only stated without integrating them within a larger picture. Therefore, a number of themes, all grounded in theoretical frameworks, have been chosen, and experimental evidence is discussed below in relation to these themes.

The choice of themes to be discussed was difficult because the amount of existing research with a cognitive spirit on advanced mathematics topics is still limited, and thus it is not yet completely clear which themes are the most relevant and in what way. The choice has therefore been dictated by three factors: the apparent relevance of the themes to the investigation of advanced mathematical thinking, the number of papers presented at recent PME conferences that made explicit reference to the themes, and the personal preferences and views of the authors of the chapter. The themes that were chosen are *concept images and concept definition, process and object,* and *visualization.*

Concept Image and Concept Definition

One of the frequently recurring themes for discussion in our group concerns the difference between the "mathematical concepts as formally defined and the cognitive processes by which they are conceived" (Tall & Vinner, 1981, p. 151). To describe ensuing conflicts, Vinner has coined the terms *concept definition* and *concept image*, which may be briefly characterized as follows:

> All mathematical concepts except the primitive ones have formal definitions. Many of these definitions are introduced to high school or college students at one time or another. The student, on the other hand, does not necessarily use the definition when deciding whether a given mathematical object is an example or nonexample of the concept. In most cases, he or she decides on the basis of a *concept image*, that is, the set of all the mental pictures associated in the student's mind with the concept name, together with all the properties characterizing them. . . . The student's image is a result of his or her experience with examples and nonexamples of the concept. Hence, the set of mathematical objects considered by the student to be examples of the concept is not necessarily the same as the set of mathematical objects determined by the definition. If these two sets are not the same, the student's behavior may differ from what the teacher expects. To improve communication, we need to understand why it fails. (Vinner & Dreyfus, 1989, p. 356)

For example, students in a service level calculus course were given the graph of $y = x^3$. Most of them stated correctly that it is possible to draw exactly one tangent to the curve at the origin, but fewer than 20% were able to draw that tangent correctly (Vinner, 1982). Their concept image did not include a horizontal tangent at a point other than a maximum or a minimum. In spite of this limitation, the students had successfully used their existing concept image in a considerable number of situations. This finding is typical for all incorrect or incomplete concept images: Its successful use strengthens the students' confidence in it. Incomplete concept images are not specific to advanced mathematics; their role in geometry learning--in particular, in the formation of prototype examples and their relation to concept images--is discussed in the chapter by Hershkowitz in this volume. Cornu (1981) has studied the implications of this phenomenon for the learning process; he pointed out that such a concept image turns into an obstacle when the student is faced with a situation where, because of its incompleteness, the concept image is insufficient. In other words, the student has a point of view that is too narrow, too exclusive, and thus inappropriate for dealing with a given situation or for solving a given problem. A cognitive conflict results whose solution necessitates a change of point of view about the concept. The student thus constructs knowledge dialectically by progressing through a series of concept images whose evolution is conditional on overcoming cognitive obstacles.

It might be useful to point out here that the theoretical constructs of concept image and of obstacle, which were developed by different researchers and represent different theoretical frameworks, complement each other in the explanation of phenomena occurring during the learning process.

Processes and Objects

P. W. Thompson (1985) has proposed a theoretical framework in which mathematical knowledge is characterized in terms of processes and objects. His ideas are based on and extend Piaget's, mainly the notion of reflective abstraction. Similar theories have been proposed by other authors (Douady, 1985; Dubinsky & Lewin, 1986; Sfard, 1987). We give here a simplified description, mainly following Thompson but taking other ideas into account.

Mathematics deals with numbers, variables, functions, and so forth, all of which may be considered as objects. These objects are connected by relationships; they are parts of structures of objects. Processes are composed of operations on these objects. They thus transform the objects. Structures may or may not be preserved under these transformations. As an example, a function may be considered as a process that associates objects from the codomain with objects from the domain. Often, these objects are numbers. Any specific function may be considered as a process that operates on these numbers: It transforms them into other numbers. There are relationships between the numbers; for instance, order. Thus the set of numbers is structured. Some functions preserve this structure (increasing functions), and others do not. For an additional example of structure, consider distance on the number line (which imposes a metric). Functions of the form $f(x) = x + b$ preserve this structure; functions of the form $f(x) = ax + b$ $(a \neq 0, a \neq 1)$ alter it only inessentially; nonlinear functions do not preserve it. As the student continues to progress through mathematics, he or she soon reaches a stage where it is insufficient to consider a function as a process, operating on numbers. The student encounters processes, such as differentiation, that operate on functions. It thus becomes necessary to consider a function as an object to be operated upon.

One reason for the complexity of mathematical knowledge is that most mathematical notions may take the role of processes or of objects, depending on the problem situation and on the learner's conceptualization. Typically, learning about a concept includes many stages (that may be spread far apart in time), starting with carrying out the operations of a process in concrete terms. As the learner becomes more familiar with a given process, the process takes the form of a series of operations that can be carried out in thought; the learner has achieved *operational thinking* with respect to this concept. At a further stage, the mental picture of this process crystallizes into a single entity, a new object. Once this is achieved, the student is able to think of this notion either dynamically as a process or statically as an object. This enables the student to think in terms of possibilities: What would happen if I did or did not carry out a certain operation? In these terms, one of the most essential steps in learning mathematics is *objectification:* making an object out of a process. And one of the main aims of the curriculum is to develop operational thinking, thinking about a process in terms of operations on objects.

This attempt to describe mathematics learning in terms of the construction of mental objects and processes by the learner is systematic but gives rise to many questions: Can learning and the development of understanding be described so systematically? Is the description too rigid? Is the mathematical content being lost behind processes and objects (or are they the mathematical content itself)?

Visualization

Many concepts and processes in mathematics can be tied to visual interpretations; that is, visual models can be built that reflect (a large part of) the underlying mathematical structure. Visualization, from the point of view of mathematics education, includes two directions: the interpretation and understanding of visual models and the ability to translate into visual images information that is given in symbolic form. In addition to this aspect of coding and decoding, the direct processing of information in visual form may take on central importance in learning mathematics. The pedagogical and didactic potential and problems that arise from the possibility of visual modelling in mathematics education have recently been the subject of much attention. (For curriculum related work on visualization, see the collection of papers edited by Eisenberg & Dreyfus, 1989; in particular, the review article by Bishop, 1989). This surge in interest is due to the considerable increase in the capacity of visual representation because of the growing use of computers in the classroom. Issues that are being investigated are whether (and which) concepts, objects, or processes should be introduced through one of their visual settings or in a symbolic manner? What is the effect of a visual versus a symbolic approach on the concept images that the students form? What is needed for students to coordinate the visual and symbolic representations of a given object or process? What are the drawbacks of the inaccuracy that is necessarily involved in any visual representation and of the fact that infinite structures can be only incompletely represented? And most importantly, does the use of visual representations promote or impede abstraction?

Topics

We now turn to some central topics of mathematics instruction and discuss the research on learning these topics in the light of the three chosen themes: concept image and definition, process and object, visualization.

Functions

The function concept is no doubt one of the most important in modern mathematics, and some have gone so far as to call it "the keynote of Western culture" (Schaaf, 1930, p. 500). Historically, the function concept has evolved from being a means to describe quantitative dependence relationships to an abstract mathematical construct that has been generalized, even "generalized out of existence" and used as the basic, primitive undefined concept in category theory much as a point is used as such an object in Euclidean geometry (Kleiner, 1989). This progressive sequence of abstractions has led through many levels, the most relevant of which for this chapter is the function as a one-valued correspondence between two sets (the Dirichlet-

Bourbaki definition), a definition often used nowadays in high school and even junior high school curricula. In other words, the teaching of the function concept has undergone a development analogous to the historical one; the progressive abstraction of the concept in mathematics has found its reflection in the school curriculum. The function concept became a concept with many layers of complexity and with many associated subconcepts. As a result, it may well be one of the most difficult concepts to master and teach in all of school mathematics.

Thomas (1975) was among the first to investigate students' acquisition of the function concept. He worked with above-average 11- to 14-year-olds, using a questionnaire and individual interviews. He considered real-valued functions of a single real variable and hypothesized that students acquire the function concept in stages. At the lowest stage, a function is considered as a procedure of assignment; later, a function can be identified in various settings, and translations between these settings can be performed; at the highest stage, the function becomes an object that has properties and that can be operated on, for instance, by transformations or composition. Although the verification process for the stages was only partially successful, Thomas identified a number of conceptual learning difficulties, among them using the ordered-pair point-graph representation to find images, preimages, domain, and range and performing operations on functions; in particular, the composition of functions. These results are significant from a cognitive point of view. Students may not progress in jumps from stage to stage, and it may not always be possible to identify a particular student as being at a specific stage, but the basic idea is certainly valid: Students initially view functions as procedural, as "telling us what to do to x." Later they establish connections between representations of a function in several different settings, some of them visual. Eventually, some of them reach an abstract conception of the notion, on the basis of which operations on functions such as differentiation become possible. The difficulties that Thomas identified occur exactly at the two most essential junctures in this development: visualization, that is, building the link between a visual/geometric and an analytic/algebraic setting; and objectification, that is, the transition to the conception of a function as a single mathematical entity.

Difficulties in identifying (pre)images in various settings have also been found by Markovits, Eylon, and Bruckheimer (1986). The same study showed another interesting effect, namely that junior high school students were prepared to forfeit other simplifications and even create inconsistencies in order to give a linear rather than a curved example for a function. This effect has been called *linearity boundedness* and was observed repeatedly. Karplus (1979) asked students to make interpolations between paired values given in a table that exhibited clear nonlinearity. The values came from real-world contexts such as bacterial population growth. Most high school students used linear interpolation in graphical or algebraic settings. Ponte (1984) concluded that many of his 16-year-old students made extensive use of a linear conceptual structure and that nonlinear situations are often thought of in linear terms (see also Rizutti & Confrey, 1988). Dreyfus and Eisenberg (1983) similarly asked college students to find a function through three given noncollinear points. Although most students gave a correct answer when the question was asked in a graphical setting, many did so by giving a partially linear function. When the question was asked in an algebraic setting, several students gave three linear relationships as an answer, one through each point. These students gave different answers in different settings. Again, students appear to fail to establish a correct connection between the different settings, and they perform better in the visual/graphical setting than in the analytic/algebraic one.

The transition between these two settings is a source for difficulties for students and a testing ground for researchers. The difficulties occur not only within but more significantly between the different settings--that is, in the visualization of the graph of a function given by a formula and in the interpretation of graphically given information. Already at 14 years, students seem able to handle intuitively information in tabular, graphical, and pictorial form (Dreyfus & Eisenberg, 1982). It appears, however, that this ability is on a rather algorithmic level. In three different tasks, Ponte's (1984) 16-year-olds tended to think of continuous variables in terms of discrete states, represented the variation of continuous phenomena by a set of unconnected points on a graph, and used dot-by-dot strategies to represent continuous variation. Similar observations led Janvier (1982) to suggest that distinct conceptual structures are involved in handling the same task for a function defined on a discrete domain and for a similar function defined on a continuous domain, and that no transfer occurs between them. These difficulties in the transition between different settings become much more pronounced when the tasks require considering functions as objects, as in tasks on function transformations such as shifts, $f(x) \rightarrow f(x) + k$ or $f(x) \rightarrow f(x + k)$, and stretches, $f(x) \rightarrow k f(x)$ or $f(x) \rightarrow f(kx)$ (Lowenthal & Vandeputte, 1989). Above-average ability 17-year-old students were given tasks such as identifying which of the parameters a, d, e in $y = a(x - d)^2 + e$ were identical for three parabolas given in an accompanying graph. Although the results were very poor at the outset, a short, visually based instructional unit on function transformations led to an encouraging improvement in performance. Similar results were obtained by Sfard (1989). She observed that her students acquired a procedural (or operational) conception of function long before they were able to pass to a structural (or object) conception of function. She found that the passage to a structural conception is very intricate and that therefore high school students achieve it with difficulty, if at all.

A study on the question of whether students can draw analogies between graphical and algebraic representations has shown that they cannot unless they are explicitly taught to operate in both representations (Schwarz & Bruckheimer, 1988). However, a curriculum making a systematic and concerted effort to not only present functions, from the outset, in several representations but also establish very explicitly the links between these representations does lead to a considerable amount of transfer between representations during problem solving, and this transfer is correlated with superior problem solving performance (Schwarz & Dreyfus, 1989).

Moreover, and in spite of curriculum development efforts aimed at teaching the function concept in several representations and at a fair level of abstraction, high school seniors and even beginning college students have been shown to have a very limited mental image of what a function is (Vinner, 1983; Vinner & Dreyfus, 1989). In fact, many students react only to the external, symbolic aspects and must thus be said to have no idea at all of functions. Most of the others do have a concept image that works in a limited number of cases, namely that a function is a formula, an algebraic expression. A slightly more general rule like "add two if x is negative, subtract three if x is nonnegative" is not acceptable to them. Only a few students conceive of a function as a transformation, an operation, or even a more general correspondence between sets in the Dirichlet-Bourbaki sense. It is interesting to note here that the same studies showed a serious discrepancy between what the students stated a function was (their concept definition of a function) and what they actually used when solving problems (their concept image). A student may very well recall the definition of a function as a mapping or correspondence and say so when asked, but proceed in a way that is inconsistent with or even contradictory to this general definition when asked to identify whether certain mappings are functions.

In the section on educational implications (p. 130), we discuss some curricula that give an active role to the graphical representation, and we surmise that direct visual processing helps students form more complete concept images of functions.

In summary, cognitive research on the function concept has exhibited three major interrelated problem areas:

1. The discrepancy between the mathematical definition of the concept, the concept definition that the student knows, and the mental concept image that the student actually uses when he or she is processing a problem;

2. The difficulties students have in graphically visualizing various aspects of a function and, vice versa, in interpreting information given in graphical form; and

3. The difficulty of going beyond considering a function as a procedural rule and conceiving it as a single entity, a mathematical object.

Calculus

Beginning university students are notorious for arriving in their calculus classes with far less knowledge, skill, and understanding than their instructors assume. In France, for example, a large-scale study (Robert & Boschet, 1984) has shown that whereas beginning students are reasonably competent in algebra, they have difficulties in the numerical domain (infinite decimal expansions, upper and lower bounds) and in logic (interpretation and manipulation of statements that include quantifiers). Moreover, they are generally poor in graphing, both in producing and in interpreting.

These weaknesses find their expression in cognitive research on conceptions of beginning calculus students. Several studies specifically concerning the concept of limit have been carried out. These studies have repeatedly identified the same persistent errors and the same difficulties across students in several countries and of different ability ranges. For instance, students find the distinction between limit and bound difficult; they make statements such as, "The limit of $9/10 + 9/100 + \ldots + 9/10^n$" as n approaches infinity equals 1, but 0.999... is strictly smaller than 1." And some offer a related statement: "The terms of a sequence get closer to the limit but never actually reach it"; none of the terms of a sequence can therefore be equal to the limit, and the sequence 1, 1, 1, . . . is not a sequence that tends to 1. Moreover, the variable nature of the running index in a sequence is often lost on students: They write statements such as "$\lim_{n \to \infty} u_n = 1/n$"; if n occurs twice they tend to decouple the two occurrences as in "if $\lim_{n \to \infty} u_n = 1$ then $\lim_{n \to \infty} (u_n)$" = 1." Other difficulties are of a logical nature and connected to the use of quantifiers, which are necessary in treating limits formally or even in understanding simple statements like "For every $\epsilon > 0$, $|x - y| < \epsilon$." A majority of first-year university students interpret this as meaning that x and y are not equal, but they are infinitely close. In other words, there are serious difficulties with formalization--in particular, quantification.

There are several ways of interpreting these student difficulties. Tall and Vinner (1981) used concept images and concept definitions and concluded that inappropriate concept images can seriously hinder the development of the formal theory in the mind of the student. Robert (1982) analyzed a sequence of mental models students use for dealing with limit problems. Although these models may well be seen as instantiations of concept images, they have also been shown to be correlated with student performance on limit problems, the "best" model being one that combines dynamic (changes over time) aspects and static (intervals around the limit point) aspects of the limit concept. Cornu (1981) and Sierpinska (1985), on the other hand, view the student's evolution in learning the limit concept as an obstacle course, the obstacles being incomplete conceptions that have been successfully applied in the past. More specifically, Sierpinska has proposed a list of five types of obstacles: obstacles linked to the *horror infiniti* (see below); obstacles linked to the notion of function (e.g., the lack of distinction between limit and (upper) bound may be interpreted as lack of distinction between a function and the set of its values); geometrical obstacles (what does it mean that a secant is "arbitrarily close" to the tangent?); logical obstacles (such as lack of appreciation of the role of quantifiers, as in the example above); and obstacles linked to the symbolism associated with limit processes. The most important of these five types of obstacles is the so-called horror infiniti. Examples such as the ones above concerning the variable nature of the running index and the decoupling of the two occurrences of the running index are interpreted as an unwarranted extension of the rules of ordinary algebra to the algebra of limits. Another aspect of horror infiniti is the student's refusal to consider passage to the limit as a proper mathematical operation. This refusal becomes quite clear from the manner in which Sierpinska's students expressed themselves about limits and in which they established them--for instance, by incomplete induction.

For the concept of limit, as for the function concept, the integration of information, the coordination of several different aspects of the same mathematical situation, is one of the central stumbling blocks. And again, the most serious difficulties appear at the transition from viewing a limit as a process to viewing a limit as a mathematical object, with which it is possible to operate.

The topics of continuity (Vinner, 1987) and tangent lines (Vinner, 1982; Tall, 1987b) have also been investigated, and results along the same general lines have been found. A topic related to limits is the one of the cardinality of infinite sets. Students' intuitions on this have been investigated by Tirosh, Fischbein, and Dor (1985). M. M. Wheeler and Martin (1988) exhibited inconsistencies between preservice teachers' explicit and implicit knowledge of infinity.

Research into students' understanding of the central processes of calculus--differentiation and integration--has shown similar learning difficulties. Students have a strong tendency to reduce the mathematics of this topic to a collection of algebraic algorithms, while avoiding graphics as well as geometrical images. As a consequence, they lack the ability to grasp the role of approximations, which is fundamental for understanding the concepts of derivative, integral, differential, and others. The students are unable, because of their algorithmic approach, to develop sound intuitions for the concepts involved. Investigators in France and elsewhere have exhibited students' tendency to stress the procedural aspects over the conceptual ones. For example, students who were asked to discuss the differentiability of $f(x, y) = 2x + 4y + y^3(\sqrt{1-\cos x} + x^2)$ immediately started computing the partial derivatives of f rather than studying the structure of the expression (Alibert et al., 1987). Students were also found to look at differentials as either purely fictional elements serving simply to denote the variable of integration, or as little bits of something, made to be piled one on top of the other (Artigue, 1986). Students do not see differentials as approximations, nor do they see them as functions, not even of a single variable. This research, as well as parallel work by Artigue and Viennot (1987) has led to the following conclusions about first-year calculus students: Their geometric images are very poor; their functional thinking is weak; they can compute derivatives but cannot work with linear approximations (nor do they conceive of a derivative as an approximation); and they compute primitives rather than integrate. To illustrate this last point: A group of students who were perfectly able to find primitives of given functions and to interpret them in terms of areas under curves were given the following problem:

> Find the (gravitational) force between a thin uniform bar of given length lying on a line and a point mass placed on the same line.

Very few of these students recognized that this was a typical integration problem (Alibert et al., 1987). These results are both parallel and causally related to students' lacking understanding of the function concept, to their inability to conceive of a function as an object. The parallelism lies in the fact that in both cases visualization is weak or lacking altogether. The causal link is twofold: Because of their limited concept of function students are unable to see a differential as a function; it does not fit their concept image of what a function is supposed to be. Moreover, because a function is for them a procedure, rather than an object, they do not have the means to understand differentiation and integration as operations. An operation needs to operate on something; in the case of differentiation and integration, the object to be operated on is a function. If the object does not exist, how is one to imagine what the operation is or does?

However, the problems with the central concepts of calculus reach far deeper; although students may be able to find primitives, they are far from understanding the mechanisms of approximation that underlie the theory of integration, such as locally approximating the function to be integrated by a piecewise constant function. In order to understand this mechanism, a student needs not only to be able to operate on a given function but to operate in the abstract on a nonspecified function and to formulate approximation processes such as Riemann integration in terms of such nonspecified objects.

The central question from the point of view of advanced mathematical thinking is the following: To what extent and how do students form concept images and construct the abstractions necessary for the kind of mathematical thinking characterized as advanced in the next section? In summary, the research discussed in this section has shown that:

1. Students learn the procedures of calculus (finding limits, differentiation, etc.) on a purely algorithmic level that is often built on very poor concept images. Difficulties with a process conception can be explained in terms of the students lacking the necessary high level of abstraction (objectification) of both the function concept (as an object) and the approximation processes involved in differentiation and integration.

2. Visualization is rare, and if it occurs the cognitive link between the visual/graphical and the analytic/algebraic representations is a major point of difficulty.

Mathematical Behavior

Most of the research on specific advanced mathematical topics can be discussed according to the three chosen themes in a manner similar to the discussion of the research on functions and on calculus. There is, however, a body of work that is very relevant for advanced mathematical thinking without being related to a specific mathematical topic. This body of work cuts across topics and the themes are underlying it. It is concerned with the social and psychological aspects of "behaving like a mathematician." It deals with such general issues as problem-solving behavior; defining, conjecturing, and proving; elegance and aesthetics; and modelling and applicability. According to Michener (1978), these are among the mathematical *know-hows* that every mathematician possesses but that are not explicitly or consciously taught by university level instructors: such know-hows as how to construct a proof or a counterexample, how to select a generic instance, or how to sharpen a definition. These know-hows are not part of the overt content of any undergraduate course, yet without some understanding and appreciation of them, students are doomed to a low-level imitation of the mathematics they are shown by their instructors. In a seminar with six mathematics majors, at the Massachusetts Institute of Technology, Michener successfully brought these know-hows to the surface and encouraged her students to discuss their application to the subject matter. Although many of these hidden features of behaving like a mathematician are of interest, on only a few of them has research work been systematic and extensive; two areas are typical: a body of studies concerning proof and investigations on problem solving.

Proof

It is not accidental that proof is the one mathematical know-how for which enough cognitive research has been reported to make at least some tentative conclusions feasible. Proving is one of the central characteristics of mathematical behavior and probably the one that most clearly distinguishes mathematical behavior from scientific behavior in other disciplines.

Several studies have contributed to a clarification of students' views on what a proof is. The nature of proof is very nebulous even to many college students, as evidenced for example by the student who compared 1 and 0.999... by stating that 1 is larger but it can be proved that they are equal (Cornu, 1981). They also seem to remain unconvinced that a single counterexample invalidates a statement (Alibert, 1988b). Although there is a tendency in many countries to postpone the time when students must master the techniques of formal mathematical proofs, researchers have found it instructive to study the informal origins of proofmaking in the work of younger students. In a seminal study, Bell (1976a) asked 15-year-old students to prove some elementary but nonstandard propositions. He found that 18% of the students made no sensible response; 25% tested a few numerical examples and accepted this as evidence constituting a proof; 35% attempted, though unsuccessfully, to appeal to some general principles; and 21% produced an approximation to a proof, though with some omissions. Analogous research questions were addressed by Fischbein and Kedem (1982). They tested 15-, 16-, and 17-year-old Israeli high school students and found that even when the students had been given a formal proof of a proposition and accepted it as valid, at least two thirds of them felt that their conviction about the correctness of the proposition would be strengthened by further empirical checks. Similar results were found by Porteous (1986) among 11- to 16-year-olds in England and by W. G. Martin and Harel (1989) among preservice elementary teachers in the

U.S.A. These prospective teachers received extensive and explicit instruction about the nature of proof and verification in mathematics. Nonetheless, many accepted inductive arguments as proofs of mathematical statements to about the same extent as they accepted deductive statements.

Balacheff (1988a, 1988b) carried out an extensive study with 12- to 15-year-old students. He identified similar difficulties and then went on to study the types of arguments by means of which they satisfy themselves of the truth of a proposition. He thus classified proofs into *pragmatic proofs* and *intellectual proofs*. Pragmatic proofs include naive empirical arguments and arguments based on crucial experiments. When generating this kind of proof, students essentially assert the truth of a statement on the basis of observed facts. Intellectual proofs include arguing from generic examples and using thought experiments; they assert the truth of a statement based on some process of reasoning, usually the formulation of a general argument. Balacheff investigated the transition from pragmatic to intellectual proofs and the development of students' proving abilities in the social situation of the classroom. He concluded that although social interaction can act as an incentive to the development of efficient argumentation, it may also constitute an obstacle to the development of mathematical argumentation. That happens, for instance, in the case of counterexample: Whereas an objection to a claim usually only casts some doubt on the claim and allows for further argumentation, a counterexample to a mathematical statement invalidates that statement irrevocably. The didactical conclusions from Balacheff's work are that the responsibility for establishing whether statements are true or false should be transferred from the teacher to the students and that proving should be learned together with the mathematical content rather than as a separate topic. For Balacheff's 12- to 15-year-olds, a proof became a mathematical tool, a process being used but not explicitly considered as an object in its own right.

Balacheff's proposals have been put into practice in several places. Hadas, Dreyfus, and Friedlander (1983) developed and implemented a Euclidean geometry course in which students were led to deal explicitly with the logical ingredients of a proof--for example, that a statement is correct only if it is correct in every case. One of the principal techniques they used was to present students with a statement and to let them find out and discuss whether it was true or false. Similarly, the distinctions between assumptions and conclusion and between a theorem and its converse were explicitly addressed. Hadas et al. found that students who took their course improved more than others in their ability to justify conclusions and to refute wrong statements by counterexamples. Alibert (1988a) recently applied Balacheff's ideas at the university level. Proposing the view that undergraduate students do not understand that formal proof is a functional and powerful tool, he presented students with problem situations about which they were invited to make conjectures. The class then examined the conjectures and argued over them. The students were thus forced to become producers of their own mathematical knowledge. Similar work with mathematically uneducated adults has been described by Drouhard, Lymberopoulou-Fioravantes, Nikolakarou, and Paquelier (1988). Their teaching approach was aimed at turning the teacher-student communication from a valuation of "good" versus "bad" to one of "true" versus "false." They did this by allowing the students to discuss problems rather than simply solving them in order for the teacher to judge the student. They generated opportunities for the students to express meta-knowledge as well as knowledge (Schoenfeld, 1985; see the section on problem solving below). Then they caused the students to commit themselves in discussions about the truth of their statements and to produce supporting arguments (see Dreyfus, in press, for an account of this and other work with adult students on justification and proof). In the teaching experiments described here, the differences and the relationship between conjecturing processes and deductive processes becomes crucial. This relationship has been investigated in the context of geometry teaching and is discussed in detail in the chapter by Hershkowitz in this volume.

The studies described above focused on the question of what a proof is (and how to make one). Proofs in advanced mathematics are often complex, and the need arises to focus on the question of what is *in* a proof. This question has been explored in several recent studies. Dubinsky and Lewin (1986) analyzed the conceptual components that students need in order to understand mathematical induction and the connections between these components that they need to establish in the process. Dubinsky and Lewin found that students first need to know methods of proof (e.g., by contradiction), functions (in the sense of algebraic expressions into which numbers can be substituted to obtain a result), and logical necessity (the existence of situations in which B follows from A with absolute certainty). On this basis students need to construct mentally the notion of an implication $A \to B$ as an object and to generalize their function concept to implication-valued functions. These notions must be coordinated with modus ponens, a further development of logical necessity in order to generate the concept of induction. According to Dubinsky and Lewin, students who fail at any one of the above junctures find themselves unable to grasp fully the idea of a proof by induction. A similar decomposition as for induction has been under-taken for another topic that is basic for appreciating proofs, namely quantification (Dubinsky, 1988). Sfard (1988) has also analyzed proof by induction but from the process versus object point of view. She has found that, as for the function concept, a process approach is easier and increases the likelihood that students will grasp the principle of induction.

Leron (1985) makes the point that proofs should not be presented as a linear chain of argument but according to their structure. He analyzed a number of proofs (Leron, 1983b) and concluded that all nontrivial proofs appear to be amenable to a top-level description in a few lines. This description is based on the construction of what Leron calls the *pivot,* a mathematical object that establishes a direct, easily grasped connection between the premises and the conclusion. This provides an overview, a top-level view of the proof. Then the pivot needs to be dealt with, usually in the form of a system of constraints that needs to be solved. This may be complicated and involve much detailed work, but it can often be done recursively by the same structuring technique as the entire proof. As an example, consider the proof of the statement that the limit of a product of two functions is the product of the limits: If $\lim_{n \to a} f(x) = L$ and $\lim_{n \to a} g(x) = M$, then $\lim_{n \to a} f(x)g(x) = LM$. Here the top-level description of the proof is only that for any given $\epsilon > 0$ one can find a $\delta > 0$ such that $0 < |x - a| < \delta$ implies $|f(x)g(x) - LM| < \epsilon$. The pivot on which the proof hinges is the δ we have to find for the given ϵ. Searching for the δ leads to the decomposition of $f(x)g(x) - LM$ into the form $(f(x) - L)(g(x) - M) + M(f(x) - L) + L(g(x) - M)$ (or a similar one). The connection between the top-level description and the necessary estimations is established by this decomposition. The detailed estimations are not central to the proof and thus belong to a low level of description (see Leron, 1983b, for details).

In summary, in a first group of studies with high school students three categories of difficulties with proofs have been identified: a lack of the sense of a need for proof, a failure to grasp the nature of a proof as something providing logical inescapability, and difficulties in writing proofs. On the basis of these analyses, techniques of presentation have been proposed to improve students' views on the nature of proof and thus their ability to see the need for proof and to construct proofs. First attempts to implement these proposals yielded promising results for the case of simple proofs. A second group of studies has dealt with more complex proofs such as inductive proofs; structural analyses have been made and ways proposed to take the structure into account in the teaching process. There is a basic difference between the two groups of studies: Whereas in the first the students are assumed to look at a proof as something you do, in the second they are supposed to look at a proof as a piece of mathematical knowledge. In this respect there is a parallel to other topics discussed above: That is, there is a transition between proof as a process and proof as an object. No studies have been carried out that investigate the achievement of this transition. In fact, such an investigation would probably be limited to mathematics majors at university level.

Problem Solving

A widespread view holds that mathematical behavior is best investigated in problem-solving situations. Here we follow Schoenfeld (1985) in defining problem-solving situations as ones where the solver does not have easy access to a solution for solving a problem but does have an adequate background with which to make progress on it. In the section on proof, we have already pointed to the fact that discussion about mathematical statements and problems plays a central role in realizing the need to justify and in judging what constitutes a valid justification. The importance of such metacognitive aspects for mathematical behavior has been particularly well realized in research on problem solving. The state of the art in research on mathematical problem solving for all age groups and all levels of mathematics is well reflected in the volume edited by Silver (1985). Although this work is important for problem solving in general, Schoenfeld's work is particularly relevant to problem solving in advanced mathematics and is therefore described here as representative. It is based on Polya's classical analysis of problem-solving activities but goes beyond Polya's ideas in taking into account not only general problem-solving strategies but also local ones, which are related to the mathematical topic under consideration, as well as the interplay between global and local strategies. Schoenfeld investigated students' mathematical behavior while they thought about problems with substantial mathematical content. He asked questions such as how the problem solver decides which mathematical knowledge to access, how to use it, and what the relationships are between these choices and the solver's understanding of the area of mathematics involved. The questions are thus questions of heuristics and control in problem solving but also questions of resources--in particular, the mathematical resources available--and of beliefs about what these resources are useful for. Schoenfeld has developed research methods for observing students during problem-solving sessions in such a way as to allow the researcher to glean information relevant to these complex questions. Some of the beliefs he has identified in high school students are that form in mathematics is more important than substance, that doing mathematics means "going through the steps" according to predetermined rules, and that all mathematical problems are solvable in a few minutes or not at all. It is not difficult to see that these beliefs follow naturally from the instruction the students receive (Travers, 1986). Control, for instance, which is essential for success in solving problems as defined above, is usually not the students' responsibility in typical exercise situations. In a series of studies carried out within the framework of problem-solving courses, Schoenfeld has also shown that mere problem-solving experience is not enough to remedy the situation. Heuristic strategies, planning solutions, and recognizing the deep structure rather than the superficial properties of problems are skills that have to be taught explicitly. If they are, then some changes come about. Rogalski and Robert (1988) have shown that this is true even for students who are rather advanced in their mathematics studies: high school seniors and college science majors. Explicit instruction in local and global problem-solving methods for complex geometrical problems and problems dealing with the convergence of numerical series was efficient and had clear and rapid effects not only on students' problem-solving ability but also on their beliefs. Students have been observed to apply these high-level skills on their own initiative, and thus to improve their problem-solving performance. In other words: Problem solving can be taught and learned.

Educational Implications

Students who participate in courses on precalculus and calculus are usually able to learn the procedural aspects, the algorithms they are taught, especially within an algebraic/analytic formalism. Thereby, they tend to construct incomplete concept images that are too closely linked to the manipulations they have learned; they are often unable to establish connections. Visualization in the sense of building a geometric model for an analytically given situation or vice versa is virtually nonexistent. This deficiency prevents students from establishing the conceptual links necessary to see the abstract mathematical structure hidden behind the concrete realizations of a concept. In other words, concepts seen by mathematicians as objects--as single, structured entities--are seen by students as an assembly of loosely connected procedures with symbols. Abstraction is sorely lacking.

Many of these difficulties can be explained by classroom practice. Teaching in high schools and colleges tends to stress algorithmization of an algebraic type to the extent that students faced with a convincing visual argument typically react with "Okay, but can't you give a mathematical reason?" Not only does common teaching practice give a low priority to visual/geometric aspects of analysis, but where these aspects are present, they are only loosely connected to the algebraic argument. Students are not usually asked to construct the bridges, the links.

Gardiner (1982) has described the emergence of the function concept as a tug of war between two mental images, the algebraic (a formula) and the geometric (a curve). This war should be settled by considering the concept as an abstract notion whose concrete representation may be algebraic, graphical, or any of a number of others. Naturally then, the question arises as to how to teach such an abstract and complex concept. How do students understand functions in different representations, and more importantly, how are they able to make the transition between representations? If a constructivist approach to building curricula is taken (P. W. Thompson, 1985), then activities explicitly enforcing these transitions are a prerequisite for the formation of concepts in the sense of the objectification referred to above. Visualization, or more precisely the bridges between visual and analytic representations of the same mathematical concept, may thus form a powerful agent on the way to abstraction.

Computer environments are an ideal tool to support curricula implementing this line of thought. Several comprehensive developments have been undertaken (Artigue, 1987; Schwarz, 1989; Tall, 1985, 1986). Schwarz has developed an introductory functions curriculum based on a computer environment, the *Triple Representation Model* (TRM). This problem-based curriculum has been specifically designed to focus attention on within and between representation relationships. The student learns by operating in the algebraic, graphical, and tabular representations and by measuring and comparing the effects of an operation in various representations. The software and with it the curriculum thus have operational character. For example, a good strategy for finding the solution of the equation $f(x) = 7$, with $f(x) = x^3 - 4x + 1$, is first to build a table of values that roughly locates the solution (between 2 and 3), then to draw the graph within the corresponding bounds ($2 < x < 3$, $6 < y < 8$), and finally to compute the solution algebraically to within a certain accuracy on the basis of the approximate graphical solution. The TRM curriculum promotes such strategies.

Tall's (1985, 1986) graphic calculus is based on a cognitive approach to the curriculum. It uses software that is specifically designed to enable the user to manipulate generic examples of a specific mathematical concept or a related system of concepts and thus to grasp a *gestalt* for a whole concept at an intuitive level. The learner is directed through a suitable sequence of activities with examples and nonexamples towards the generic properties of the concept, this dynamic process helping him or her to construct an abstract version of the concept. Sample concepts are the gradient of a graph (exhibit local straightness through magnification), the area under a curve (give meaning to the sign of the area using dynamic growth), or the solution of a differential equation (see it develop graphically in real-time computer simulation).

Artigue (1987) also builds her curriculum on suitable computer software, which is used both in graphical and numerical mode. The computer is used as a tool to help the student develop a qualitative approach to differential equations. The role of the computer is essential here because it reduces the complexity of the problems without losing the global picture associated with them. It is used in the graphical mode to search for curves compatible with a given slope field; then the numerical mode is brought into play. The entire process is built on a continual process of association between graphs and equations. Graphics--visual information--are given a role beyond that of merely being another representation; they are central objects; information is processed visually as well as symbolically.

These curricula use an approach centered on processes of construction: The control of the activity is with the student, a coordinated development in both algebraic and graphical representations is produced, and the computer is used to achieve these goals. The curricula thus have the characteristics that have been recognized to be central for promoting concept development and problem solving (Dreyfus, 1984; P. W. Thompson, 1985).

These curricula have been tried in experimental setups. Tall (1986) and Schwarz (1989) have used experimental and control classes and shown the experimental curricula to be clearly superior to the standard curricula in what they were designed to do. Moreover, experimental students had the same skills as control students in classical computational tasks. Artigue (1987) has analyzed classroom learning processes in terms of a detailed a priori analysis of the teaching/learning process and has observed a large measure of agreement between the expectation and the realization.

In summary, curricula can be designed to encourage the formation of appropriate concept images in precalculus and calculus. The curricula mentioned stress process and object aspects of mathematical notions through their operational character; they use visual and analytic modes of action and specifically address the links between the different modes. These curricula appear to be successful in enhancing concept formation without neglecting skill development. Appropriate computer software is a choice technique for achieving these aims because of (a) the natural way of implementing the operational character of the curriculum and the ease with which several modes can be dealt with in parallel, and (b) the obvious advantage of the computer for presenting the visual models.

This section has concentrated on the educational implications of the research related to analysis. It has not touched, so far, what was said above about students learning to behave like mathematicians. There is, however, a strong link between the two. The implementation of the curricula described here leads to a type of activity on the part of the student that goes beyond the solution of standard exercises according to sample problem solutions provided by the teacher. The students are called upon to investigate mathematical situations by asking questions of a "what happens if" type. Thus the students have to take some of the responsibility for the action of learning. The main conclusion reached in the section on mathematical behavior was that students should be led to become producers of their own mathematical knowledge. The students thus have to assume a much heavier part of the responsibility, deciding on the truth or falsity of mathematical statements. In this sense, the curricula described above can be seen as a step in the direction of learning mathematical behavior.

Conclusion

Only a small number of trial curricula have been developed and tested so far, and a vast number of cognitive questions remain open. In fact, although specific research questions may have found an answer in specific studies, none of the larger questions has yet been answered. There are some indications from analysis that a description of mathematics in terms of objects and processes corresponds to the cognitive structures that students can be expected to develop, and that such an object/process description leads the students to develop appropriate concept images. It also appears that visualization plays an important role in this development and that open-ended learning environments are tools that are well suited for presenting mathematical topics in a manner that stresses objects and processes while using visual and analytic descriptions in parallel. Moreover, in curricula centered on such learning environments it is easy to transfer some of the responsibility from the teacher to the student, and this transfer has been shown to support concept learning.

Although all these indications are positive, the causal links are largely unknown and open a wide range of research questions. For which mathematical topics are visual models supportive of the cognitive development? What are the mechanisms mediating the establishment of links between visual and analytic representations? To what extent and in what cases do visual representations hinder cognitive development because they are an additional load, an additional representation to be dealt with? Is time a factor here? Is this additional visual representation desirable or even

necessary for promoting abstraction? Although it sounds paradoxical, does concretiza-
tion by visual support facilitate abstraction on the part of the student? How does
transferring responsibility to the students interact with their cognitive development?
What are the mechanisms that lead to the (re)production of mathematical knowledge
if responsibility is transferred to the students? Although these questions have been
formulated here in a very general way, they can and must be made specific for each
concept, for each curriculum, for each learning situation. And the answers may well
be situation dependent.

Although there are many wide open questions, there is one conclusion that can
certainly be drawn even at this early stage of cognitive research into advanced
mathematical thinking: Issues such as visualization, problem-solving strategies, and
links between representations that are known to be problematic are so because they
have been assumed to be learned automatically, whereas now we know that they need
explicit attention in the teaching/learning process. They need to be explicitly taught,
not as separate topics but throughout the curriculum--that is, within each topic in a
parallel manner. These issues can only be fruitful if they are underlying the learning
process over a long period of time.

7 FUTURE PERSPECTIVES FOR RESEARCH IN THE PSYCHOLOGY OF MATHEMATICS EDUCATION

Nicolas Balacheff
with the collaboration of
Alan Bell, Alan Bishop, Thomas Cooney, Willibald Dörfler,
Celia Hoyles, Douglas Jones, Nitza Movshovitz-Hadar,
Thomas Romberg, and Nurit Zehavi

The past 13 years of activities of the International Group for the Psychology of Mathematics Education (PME) have provided researchers in mathematics education with a wide range of valuable interactions. The fruitfulness of these interactions is partly due to the variety of these researchers' backgrounds, which has permitted confrontations and collaborations among psychologists, mathematicians, and mathematics educators; it is also due in part to the exchanges at an international level that have facilitated discussions of different research perspectives, as is clearly exemplified in several sections of this book. In particular, the willingness of PME to encourage and support the existence of working groups and discussion groups alongside traditional research reports during its annual meetings has played an especially crucial role in creating a dynamic movement leading to new perspectives in this field of research.

New fields of research, such as research on learning processes in a microcomputer environment and on the learning of mathematics at an advanced level of education have emerged. But more important is the fact that beyond the initial psychological *problématique*,[1] the debates about the presented research made clear the need to take new aspects into account. Among the new aspects we would like to mention the

---

[1] We call *problématique* a set of problems with reference to a specific theoretical framework. It refers to the criteria we use to assert that these problems are relevant and to the way we formulate them.

following, which are likely to play a major role in PME's scientific orientation in the future:

1. *The specificity of mathematical knowledge.* Research on the learning of algebra, geometry, or calculus cannot develop without a deep epistemological analysis of what the concepts considered consist of as mathematical concepts. Also, it is recognized that the meaning of mathematical concepts relies not only on their formal definition but more fundamentally on the processes involved in their functioning. It is for this reason that emphasis is put on *the study of students' cognitive processes* rather than on their skills or actual productions.

2. *The social dimension.* Both the social status of the knowledge to be learned and the crucial role of social interactions in the teaching process make the social dimension an important consideration for research. One of the main steps in the development of research in the psychology of mathematics education is the movement from studies centered on the child towards *studies centered on the student* as a learner in the classroom. The student is a child involved in a learning process within a specific environment in which social interactions with other students and the teacher play a crucial role. With this evolution of the problématique, more research has been developed that requires systematic classroom observations or the organization of specific didactical processes. *Such research requires new theoretical and methodological tools* to produce results that are robust both theoretically and also with respect to their meaning for practical purposes.

As some of these questions have already been presented in this book, we focus in this chapter on the following directions for the future development of research:

- cognitive processes and the construction of meaning;
- the social dimension of teaching-learning phenomena;
- teaching mathematics as a research field; and
- new questions related to microcomputer environments.

Cognitive Processes and the Construction of Meaning

The work done by PME members covers a vast range of studies, from counting to calculus, from basic operations of arithmetic to problem solving. As Collis (1986) reminded us, it has its roots not in general psychological theories to be applied but in content-specific questions related to learning processes in the context of mathematics teaching. Explicitly or not, most of these studies convey an interesting viewpoint that has been very well expressed by Bergeron and Herscovics (this volume) in the case of arithmetic: Mathematics learning can no longer be confined to the acquisition of skills, but rather it is perceived in terms of *thinking processes.*

This shift of focus from competency and skill towards cognitive processes is reinforced by the general evolution of our problématique from the study of *students' difficulties* towards that of *students' knowledge* that underlies them. Even the term *misconception* tends to be replaced by expressions like *students' conception* or *concept image* that acknowledge a genuine construction of knowledge by the learner in the teaching situation. All the preceding chapters make it clear that, beyond merely observing students' errors while they performed some mathematical task--that is to say, considering errors as facts that we can count and classify--we need to propose explanations of their origin. The search for such explanations has brought to the forefront of the scene a complexity that previous approaches were unable to show. As noted in the chapter by Bergeron and Herscovics (this volume),

> One of the more surprising findings in the last decade of research on additive word problems in arithmetic is that word problems that have the same logical structure and that call for the same mathematical operation are nevertheless handled quite differently by young children. (p. 46)

That could also be said for other mathematical topics, such as algebra, geometry, or calculus, and for older children.

These findings mean that mathematics as described in the mathematical literature is not sufficient to account for the complexity of an item of mathematical knowledge with respect to learning purposes; the psychological characteristics of a mathematical task cannot be reduced to its mathematical properties. Two important questions are raised as a consequence of this new approach.

- How are we to characterize a piece of mathematical knowledge with respect to the learning research problématique?
- What is the nature of the learner's mathematical knowledge? How are we to describe it?

Many attempts to answer these questions have already been made independently in relation to specific content or within the context of different pieces of research. For example, studies of additive problems have led to different hierarchies relative to a specific aspect of number or of addition and subtraction. But the problem now is to find a global framework to link together these *local* cognitive structures or hierarchies. The problem is not only that of unifying but also that of establishing the links among what are likely to be different problématiques, or even different theoretical frameworks.[2]

This problem is clearly called into question by the existence of various terms used to distinguish between the *concept* and the *outcome* of the concept-formation processes

---

[2] Paraphrasing Thomas Kuhn (1962, p. 13), we could write that being able to take no common body of belief for granted, each researcher feels forced to build his or her field anew from its foundations. In doing so, the choice of supporting observation and experiment is relatively free, for there is no standard set of methods or phenomena that every researcher feels forced to employ and explain. Mathematics education as a discipline is at such a prescientific stage of development, which is quite normal for such a new discipline. The fact that these questions are now elicited and considered shows that it is on the way toward a new stage of maturity.

in the individual's mind: concept and conception (Grenier, 1985), objective and subjective image (Hirabayashi & Shigematsu, 1987), concept definition and concept image (Vinner & Hershkowitz, 1980). But what the relationships are between the meaning of these expressions is not clear. They acknowledge the existence of a *student's knowledge* and its possible distance from a given knowledge of reference, something which is undoubtly shared by all the users of these expressions. But do these researchers share the same meaning for the expressions *concept, objective image,* or *concept definition?* Here is one of the important problems to be studied for the coming years insofar as it calls into question the epistemological nature of our theoretical framework.

So, one of the most important challenges for research in the psychology of mathematics education within the coming decade is of an epistemological nature. Epistemology is concerned both as it addresses to us questions such as "What kind of objects is mathematics about?" and as it questions researchers about their own views of the nature of mathematics. The answer can be obtained not solely from the text of mathematics as a science but rather by examining how mathematics works and how it constitutes a functional answer to problems we consider: "Problems to be solved are the real source of knowledge, and problem solving is also the criterion of knowledge acquisition" (Vergnaud, 1982b, p. 31). The meaning of mathematical concepts rests not so much in their definition as in the class of problems to which they allow one to find a solution.[3] In the same way, the nature of learners' conceptions can be traced not only from what learners state explicitly but also from the way they use them and from the class of problems these conceptions allow them to solve. From this approach come many questions, such as the following:

- How are we to characterize the class of problems specifically related to an item of mathematical knowledge?
- Among the characteristics of a class of mathematical problems, which ones determine students' cognitive processes and thus the nature of the knowledge they construct?

These questions to be studied are not only of a theoretical nature, but also of a methodological nature because of the way they relate cognitive behaviors and tasks. Here we must perhaps reconsider the suggestion from Lesh (1985) that idea analysis should begin with models (i.e., hypotheses about the structures that characterize students' mathematical ideas) and *then* create tasks to test these hypotheses.

Finally, if the present trends of research have pushed espistemology to the core of the future perspectives for research on psychology of mathematics education, new problems are also raised for psychology itself. One of the most important problems

---

[3] That is clearly shown by research done on the origin of students' concept image in the context of different topics (cf. sections of this book dealing with specific mathematical content).

to be mentioned here is that of the role of cognitive conflicts in the student's construction of mathematical knowledge. This problem is difficult because, on the one hand, construction is a personal process and, on the other, overcoming any cognitive conflict is neither a deterministic process nor a well-defined one. As a result, the intellectual construction following such processes could be some distance from what is expected--or even needed. A good example of this problem and its implication is the construction of mathematical sign systems in algebra. Such systems embrace both the formal and the pragmatic mathematical meaning: The production involves the construction by the learner of intermediate sign systems whose idiosyncratic character-istics create obstacles because of the lack of socially agreed convention. But to overcome such obstacles is part of the process of constructing mathematical meaning. The example also raises the problem of the role of the social dimension and the problem of sociocognitive conflicts and their implications for mathematics teaching. These problems we consider in the next section.

Mathematics Learning from a Social Perspective

All mathematics learning takes place in a social setting. It can range from individual learning, where the social influences are experienced at a distance through being mediated by an author's text, to group learning, where the social influences are immediate. All teachers, students, and classroom observers know that there are many social and interpersonal influences at work in the mathematics classroom. Thus it is imperative that researchers try to interpret mathematics learning from a social perspective if research is to have any validity and credibility for the classroom context. In particular, and because the social dimension involves many aspects, we must try to interpret, and theorize about, mathematics learning *inter*personally as well as *intra*personally.

This distinction is important to make in order that the complementarity between these two areas is understood by researchers. Mathematics learning in its educational context cannot be fully interpreted intrapersonally because of its social setting. Equally, interpersonal constructs will be inadequate alone since it is always the learner who must make sense and meaning in mathematics. Therefore, it is crucially important to recognize this intra/interpersonal complementarity, to explore its significance and how it influences the quality of mathematical meaning obtained by students in classrooms.

In order to pursue such research, it is necessary to develop both theoretical constructs and methodological tools. Many constructs from social psychology in general hold some interest for us, but three in particular seem promising. These we can think of as social motivation, social cognition, and the social construction of knowledge.

Regarding *social motivation*, there is one topic that has stimulated a great deal of research activity worldwide--namely, the fear of mathematics. For example, Buxton (1981) and Skemp (1979) explored the idea of goals and anti-goals (to be avoided by the learner), and their discussions of anxiety, frustration, and other emotions are very helpful to our understanding of how the classroom mathematical experience appears to students.

Another anti-goal identified in the literature is the *fear of success* construct found to be of great value by Leder (1980) in understanding why bright girls in mathematics deliberately avoid success and achievement in order to retain the respect and acceptance of their peers. Of course, this is a phenomenon to be seen not only with bright girls; it will be noticed by any mathematics teacher of adolescent children who apparently prefer not to succeed, and indeed not to try to succeed, for fear of losing the respect of their friends.

Whether the teacher-mediated mathematical goals are accepted by the students *as* positive goals or are converted into anti-goals is determined by various factors, including in particular whether the mathematics teacher is perceived as a *significant other* (Sullivan, 1955) by the student. The role of significant others in motivation, although developed within the psychiatric field, has value for us also. It does not require much observation of mathematics classrooms to begin to identify which persons in the classroom significantly affect the motivation of the other students. What we need to know more about is the *mathematical* dimension of that influence.

The area of social cognition concerns the ways in which teachers "know" students, students "know" other students, and students "know" the teacher. The *self-fulfilling* prophecy is well known of course, and what that research told us was that what is pedagogically significant about a psychological construct is how that phenomenon is perceived by the teacher. Attribution theory tells us that the way the learner perceives the phenomenon is also important, of course. Jackson and Coutts (1987) demonstrated, for example, that in pair groupings where students had knowledge of each others' tested abilities, the achievement of the pairs was strongly influenced by that knowledge: "The reduced scores obtained by high ability females working with a lower mathematics ability (male) partner suggests deference to the 'dominant role' of the male and also suggests deference behavioral success avoidance in high ability females" (p. 90).

Finally, we can see the importance emerging of the idea of the *social construction of knowledge*. Mathematical knowledge, like any other kind of knowledge, is socially constructed. This idea is particularly important because it points to the need for researchers to develop more appropriate methodological tools for this research.

In the area of mathematical tests and tasks, there has been some development in *group tasks*. Many tasks and mathematical activities appear to be of the unitary type where "a task allows for a single correct answer only, or when all students must cover the exact same material in order to reach a solution" (Sharan, Hare, Webb, & Hertz-Lazarowitz, 1980). If we are to study the *social* construction of mathematical knowledge, it is necessary to have an understanding of ranges of tasks from unitary to cooperative teamwork activities, from one-right-answer tasks to many-answer tasks.

The use of video, sound tapes, and transcripts holds a great deal of promise for analyzing social interactions, although the real challenge is how to observe, document, and describe interpersonal influences. As the number of these studies increases, we can expect improvements in both the quality of the observations and the interpretive analyses. So far the main areas of interest have been with small groups in problem-solving environments (e.g., Alston & Maher, 1988b; Joffe & Foxman, 1986; Resnick & Nelson-Le Gall, 1987; Yackel, 1987) and in computer environments (e.g., Hoyles & Sutherland, 1986), and in both cases developments are very promising.

A third aspect of methodological tool development concerns experimentation with different grouping arrangements and with the role of the teacher in relation to these. Here one can cite the research of Webb (1982), who has been undertaking this kind of work for some years, and we have already cited similar work by Jackson and Coutts (1987). These researchers feel that it is only by grouping and regrouping children experimentally and systematically that we will learn about the ways that influences and patterns begin and become established.

This field holds much promise for mathematics educators. Already we are beginning to see a growth in interest in group teaching methods, discussions, and paired work. The research and reviews by Noddings (1985) is helpful in this context. The more this research proceeds, the more teachers will become informed about ways of using these pedagogical techniques, and the richer our knowledge of how to help teachers to deal with mathematical learning difficulties will become.

Mathematics Teaching as a Research Field

To take as a research question that of students' learning of mathematics means that the child is no longer considered as an isolated system. The social dimension, as discussed in the preceding section, demonstrates that the learning process in the mathematics classroom must be understood in the context of social interaction and must take into account the specificity of the situation in which it takes place. The child appears then as a subsystem within a larger one that comprises other subsystems, such as the teacher and the knowledge itself; each reacts to the others and evolves within the teaching process. By the word *student,* we thus refer to the child as such a subsystem.

The relevance of this movement in the problématique of psychology of mathematics learning is the evidence that what teachers do in the classroom does make a difference in terms of what students learn.

The following three key issues should be considered for further research projects.

Teachers' and Students' Conceptions of Mathematics

Teachers' conceptions of mathematics and of teaching mathematics influence their instructional behavior (see, e.g., Cooney, 1985; A. Thompson, 1982, 1984). But what we know virtually nothing about is the interactive nature of teachers' and students' conceptions: how one influences the other, their origins, the intensity with which they are held, and how permeable they are in the face of classroom dynamics. These conceptions and the means by which they influence the direction and magnitude of teaching and learning constitute part of the *hidden dimension* of teaching mathematics (Bauersfeld, 1980).

This conception of the teacher is what Brousseau (1986a) refers to as *teacher epistemology:*

> The teacher is led to make explicit a method aimed at producing the answer: ways of answering by using previously acquired knowledge, ways of understanding, of elaborating new knowledge, of "applying" the teaching, of recognising the questions, ways of learning, guessing, solving, etc. In this fashion, the teacher takes as reference an implicit philosophy of mathematics or a model (like elementary geometry) specifically devised for the use it will be put to: Solving the conflicts in the didactic contract . . . . In order to teach these elements of knowledge, a teacher must therefore reorganise them so that they become suited to the description given to this "epistemology". This is the beginning of the process of modification of the elements of knowledge, a process which will alter the organisation, the relative importance, the presentation, the game . . . relative to the requirements of the didactics contract. (p. 113)

It is this process that is called *didactical transposition* (Chevallard, 1985a).

The Didactical Contract

As has often been demonstrated, many students can obtain the answer to a given question not by way of the expected mathematical reasoning but through a decoding of the didactic conventions (Brousseau, 1984; Hirabayashi & Shigematsu, 1986). This phenomenon is one consequence of the fact that in any teaching situation the teacher attempts to get students to understand what she or he expects them to do. Theoretically, passing from the teacher's information and instruction to the expected response should require the students to apply the piece of knowledge considered, whether it is being learned for the first time or is already known. But as far as it is acknowledged that to *do* mathematics is to search for and solve certain specific problems, then the teacher must bring about not the communication of a piece of knowledge but the *devolution*[4] of a problem specific to its construction. If this devolution works, the students "enter into the game," and if they win in the end, the learning has taken place (Brousseau, 1986a, p. 112; Movshovitz-Hadar, 1988).

But, as several studies suggest (Kesler, 1985; McGalliard, 1983; Owens, 1986; A. Thompson, 1982), the conceptions teachers hold may prevent them from teaching mathematics as a subject that not only is rich with respect to presenting problems but is itself problematic. For example, the fact that over 50% of the eighth-grade students in the United States (McKnight et al., 1987) think that learning mathematics is

---

[4] "A delegating of authority or duties to a subordinate or substitute," according to the 1979 *American Heritage Dictionary of the English Language.* Following Brousseau (1986a), we use this word to name the process through which the teacher "passes" to the student the responsibility for the validity of the solution to a given problem.

primarily a matter of memorizing rules suggests that students' conceptions may be similarly contrary to seeing mathematics as a field of investigation where problem solving is the main activity. Also, teachers tend to conceive of teaching mathematics as showing or telling students the "proper techniques" and helping them to reach the "correct" (often meaning the teacher's) way to think about mathematics (e.g., Kesler, 1985; Meyerson, 1978; McGalliard, 1983; Owens, 1987). They tend to rely on the external authorities of curriculum, grades, and extant mathematics to legitimize their teaching. Within such an orientation, the goal of students is likely to be one of looking to an external authority for a definition of what is mathematical and of striving to satisfy that authority with the form of an answer. In such cases, students' beliefs about mathematics and about problems are reorganized in an effort to solve problems that are more *social* than mathematical and to maintain the perceived status quo of school mathematics (Balacheff, 1982; Bauersfeld, 1988; Cobb, 1986).

So, such studies on the didactical contract and its relationship to learning processes are essential as long as what is under question is the real meaning of the knowledge constructed by the students.

Situations

Following his analysis of the social dimension, Bishop (1985) asserts strongly that "if there is one thing to be learned from research into social aspects of mathematics education it is that the *context*, and the *situation*, are all important" (p. 4). What we have developed here confirms the importance of studies taking the teaching situations as a research object: how they function with the purpose of allowing students' learning, how they affect the meaning of knowledge, and how the various subsystems interact within a given teaching situation.

As long as we consider this situation specific to the teaching of a given piece of knowledge and not in all its educational aspects, such a problématique defines a specific field of research whose object is the study of the *situations* and *processes* that have been organized with the intention of giving the students a sound knowledge of mathematics.

Within this situational problématique, questions arising directly from current practice could be studied as, for example, the following: How effective is the learning from individualized schemes using booklets or work cards compared with class or group-based teaching? There are a general awareness of a superficiality in the learning from individualized schemes and a variety of ad hoc practices aimed at alleviating the dangers, but there is no real evidence on this question.

From the point of view of studies of learning processes and phenomena, which is indeed the object of the psychology of mathematics education, the central question becomes that of characterizing the relationships between situational aspects and students' cognitive behavior. This situational dimension, which underlies--explicitly or not--any study on teaching processes, is seldom considered as a research object per se. Interactions are organized between students and a given milieu, between students themselves, or between students and the teacher, but more often than not the relationships between the characteristics of these interactions and the students' cognitive behavior or productions are not called into question or discussed (see Brousseau, 1986a, for an attempt to set up a theory of didactic situations to cope with this issue). Here is an important field of investigation for future PME research activities.

New Questions Arising From the Microcomputer Environment

Obviously, microcomputer devices are likely to play a larger part in the teaching-learning environment at all levels of schooling. Beyond the first steps, which mainly consisted of designing innovative classroom computer-based settings, some fundamental research questions arise that address the way knowledge is affected by such an environment or how and to what extent students' cognitive behavior and constructs are modified. A significant part of the programs of PME annual meetings have already been devoted to this field of research (e.g., Hoyles, 1987; Tall, 1987a). We restrict ourselves to some of the more significant questions, organizing them around the main types of computer learning environments: microworlds, professional software, and intelligent tutoring systems. Finally, we consider research on possible changes in classrooms because of the use of computer environments.

Microworlds as Learning Environments

One fundamental idea related to a microworld as a learning environment is to simulate on the computer (especially using its graphic facilities) situations in which the learner is prompted or guided to carry out actions and operations that are likely to lead to the conceptualization of genuinely mathematical concepts and operations. Among the first examples are the Logo environments for certain geometric concepts (see Hershkowitz, this volume). Other examples of such environments have been developed by Kaput, Luke, Poholsky, and Sayer (1987); P. W. Thompson (1985); Schwartz and Yerushalmy (1987); Baulac, Bellemain, and Laborde (1988); and others.

Psychological research is needed on this topic insofar as the construction of a learning environment demands sound knowledge about possible learning processes, error types, and the like, in the particular part of mathematics. For instance, facts and theories from the psychology of number concepts and operations are a prerequisite for the construction of effective learning environments through which the learner can construct adequate personal knowledge about numbers (P. W. Thompson & Dreyfus, 1988). Of course, a rather deep knowledge of mathematical content and epistemological considerations are of equal importance. Of great interest, then, is psychological research that looks at the effect of a given learning environment, investigating what and how students learn in that environment. This is a kind of evaluation that will again support the improvement of the environment itself.

Some research has already provided some evidence of the positive contribution of a microworld environment. For example, Sutherland (1987b) showed that students' experience with Logo enriched their understanding of variable in an algebra context.

Such an environment can provide a cognitive scaffolding for the learner--a context based on the interaction between symbolic and visual modes of thinking, in which concepts can first be *used* and later more properly *understood* (Hoyles & Noss, 1987a). Also, as Hoyles (1987) points out in her comment on a set of communications on geometry in computer environments, perceptual ambiguities, or conflicts, which inevitably arise from students' activities in an interactive computer environment, provoke students to be suspicious of perceptual cues, rethink their original intuitions, and try to make explicit their geometric knowledge in relation to the graphical feedback.   On the other hand, some difficulties can occur that should not be underestimated, such as the difficulty of linking algebraic and geometric representations in the context of the graphical representation of a function (Tall, 1987b). Using computers could lead to specific conceptual obstacles in the learning process. This important point, mentioned in the previous chapters, is likely to be one of the key questions to be explored in the future.

The Use of Professional Software

Cognitive processes need certain tools for their adequate execution. Such tools are, for example, language, graphical representations, concepts and prototypic representatives, symbols and symbolic operations (as used especially in mathematics), mental images, and models of material situations.   The computer appears able to offer qualitatively new thinking tools mainly through its graphic facilities and the possibility of the synchronous representation of different but related processes and situations. This power of the computer is exploited in the construction of microworlds, but there are other examples of thinking tools offered by the computer such as electronic spreadsheets, expert systems, data bases, word processing, and simulation programs (for economic and social processes). What is interesting and important is not only that these software products enlarge or broaden already existing or utilized ways of thinking but that they demand qualitatively new ways of thinking. At the same time, the opportunity is available to develop these new thinking techniques by using the software appropriately.   Much psychological research on the effect of using such packages is needed. Already some investigations into which cognitive skills will lose or gain in importance given these electronic tools have been carried out.   They analyze in particular the comparison of outcomes on tasks embedding specific knowledge constructed within different environments and in noncomputer environments (Gallou-Dumiel, 1987; Osta, 1987).

The case of programming languages calls for special attention not only because strong relationships exist between concepts involved in programming and mathematical concepts (e.g., Hausmann, 1985; Samurçay, 1985; Southwell, 1985) but also because of the specificity of algorithmic concepts (Cohors-Fresenborg, 1978, 1982; Laborde & Mejias, 1985; Rouchier, 1987). Much of the existing research in PME deals with Logo programming; few references exist that refer to other languages. Perhaps efforts should be made to consider other programming environments insofar as computer science concepts are likely to change the content of school mathematics teaching, as is acknowledged in some of the preceding chapters (see especially the chapters by Hershkowitz and by Kieran).

Intelligent Tutoring Systems

The general shift in curriculum development toward an emphasis on cognitive issues in the process of developing instructional systems provides the general background for the growing demand for software having diagnostic and predictive possibilities. Intelligent tutoring systems are expected to enable a better interaction between student and system and lead to better instruction; they consist of computer software that attempts to teach a certain unit or field of knowledge in the same way as experts on teaching this knowledge.

Therefore, part of such a system is an expert system into which pertinent expert knowledge is built. The system should include an awareness of psychological aspects of the learning of this knowledge. Here is again an incentive for psychological research that is at the same time a good chance for an evaluation of its results: The implemented theories must fit the reality of the learning processes.

There is another component of intelligent tutoring systems that demands intensified research in the psychology of mathematics education: Such a system should involve a *student model.* This means that in the course of the teaching and tutoring process (steered by the program), the computer tutor constructs a model of the cognitive structures of the learner from the student's reactions as far as they pertain to the intended learning process. Of special importance here are psychological theories concerning causes and cognitive structures of errors and bugs and of deficits and flaws in constructed concepts. Such theories can guide the computer tutor, for example, in its remedial teaching. Again, this topic offers an effective opportunity for testing the fit of the implemented psychological theories. In any case, it opens up a broad field for research and development for the psychology of mathematics education. (For a good account on this topic in general, see Wenger, 1987; Sleeman & Brown, 1982.)

Looking for Changes in the Classroom

The introduction of educational software creates pressure on teachers for a fundamental change in how they conduct the class (Ponte & Abrantes, 1987). The sociology of the classroom is affected: There is a need for new competencies and training for both students and teachers in order to face together new and challenging situations; the monitor is not one's copybook; the direction of student work is shared by the teacher and computer.

To ensure the relevance of research findings in this new domain of investigation for the psychology of mathematics education, especially for further application to classroom practice, we have to consider the possible changes in the teacher's role in a "microclassroom." There is a need for more explicit research on the nature of teacher interventions and how they affect learning in a computer environment. For example, how is it possible to guide students so they do not work randomly (which often might be the quickest way to the "result") and therefore do not grapple with the mathematical meaning embedded in the activity? How can the student be moved to adopt a more reflective or deductive approach that takes into account both global and local features of the task?

A significant change often mentioned, is that, whereas previously the teacher was busy teaching the subject and thus could not effectively watch the students learn, with the computer a teacher is likely to be in a better position to observe how students think and learn. According to Shavelson, Winckler, and Stasz (1985), we should be concerned that teachers be able to *orchestrate* the various modes of computer use in their classes for various students. For example, how can and should guidance vary in terms of level of abstraction and generalization with respect to different students? Some research has already been carried out in this direction (Zehavi, 1988). One point should not be forgotten: Such an environment promotes the individualization of learning and, thus, of the meaning attached to the knowledge constructed. But these meanings must in the end be compatible with those that are socially shared and recognized. Then the teaching settings, organized around the use of software, must allow a socialization of the knowledge constructed by the learner and its related meaning.

Fundamental Consequences for Further Research Agendas

The raising of new research questions to be studied in the coming years, especially those presented above, brings up the issue of the theoretical background and of the related methodology to be used.

At the beginning of PME's activities, much of the research relied on the current methodology for psychological studies, borrowing the theoretical framework elaborated by different psychological schools, among which the Piagetian school played an outstanding role. Even in this context, theoretical problems as well as methodological ones were encountered in trying to cope with the specificity of mathematics as the object of the learning processes to be studied. The development of a PME working group on epistemology has resulted in great attention being paid to epistemological issues in more and more research projects. It provides a good indication of this trend.

But the nature of the new questions to be considered calls for more efforts in this direction. Clearly, classical theories are not sufficient to cope with the problem of knowing and explaining the processes of learning mathematics concepts that take place in a teaching situation. In particular, we notice that very often research on "learning within the classroom" is still empirical, is "pragmatic," or has a very tenuous theoretical background. Such works are very fragile when confronted with practice or discussed in a scientific debate.

Significant progress has been made over the past decade, as Romberg (1981) witnesses: "We are near a significant turning point in the history of research in mathematics education. We have gone from a variety of atheoretic, fact collecting studies to a period of model building" (p. 358). In the context of this evolution we have to address fundamental questions such as the following: What is a *result* in the field of research in mathematics education? What guarantee can be given about the *validity* of these results? In other words, what are the *foundations* and *methodology* of our field of research?

Taking teachers' and students' conceptions into account has implications for our research methodologies. We must give up the notion that what we observe is in some sense isomorphic to what the observed individual experiences. We must even give up the notion that what an individual expresses is necessarily a reflection of his or her beliefs--some beliefs may be so deeply embedded in an implicit world view as to be unarticulatable. The collection, interpretation, and analysis of data must become interactive processes informed by both researchers and individual informants. Such research will require us to examine and reexamine what it means to understand one another's conceptions, to be scientific in our research, and, finally, to engage ourselves in the reflective process of examining our own research agenda and paradigms.

To borrow theoretical concepts from other fields of research, such as sociology, psychosociology, information science, and so forth, is not sufficient to answer these fundamental questions. It is time to think about the necessity for and the way to look for "homemade" theoretical concepts that will allow us to make progress in our specific field of research. PME as a group should take that question as one of its major concerns. Since PME is a group that encourages real interactions between experimental research and theoretical questioning in the field of psychology of mathematics education, it is therefore likely to support the most relevant and effective progress in this direction.

References

Alibert, D. (1988a). Towards new customs in the classroom. *For the Learning of Mathematics, 8*(2), 31-35.

Alibert, D. (1988b). Codidactic system in the course of mathematics: How to introduce it. In A. Borbàs (Ed.), *Proceedings of the 12th International Conference for the Psychology of Mathematics Education* (Vol. 1, pp. 109-116). Veszprém, Hungary: Ferenc Genzwein, OOK.

Alibert, D., et al. (1987). Le thème "différentielles": Un exemple de coopération maths/physique dans la recherche. *Actes du Colloque GRECO Didactique du CNRS, Sèvres, France* (pp. 7-45). Grenoble, France: La Pensée Sauvage.

Allardice, B. (1977). The development of written representations for some mathematical concepts. *Journal of Children's Mathematical Behavior, 1*(4), 135-148.

Alston, A., & Maher, C. A. (1988a). Effectiveness of small group problem solving within classroom settings. In M. J. Behr, C. B. Lacampagne, & M. M. Wheeler (Eds.), *Proceedings of the 10th Annual Meeting of the North American Branch of the International Group for the Psychology of Mathematics Education* (pp. 206-213). DeKalb: Northern Illinois University.

Alston, A., & Maher, C. A. (1988b). The construction of arithmetic structures by a group of three children across three tasks. In A. Borbàs (Ed.), *Proceedings of the 12th International Conference for the Psychology of Mathematics Education* (Vol. 1, pp. 117-124). Veszprém, Hungary: Ferenc Genzwein, OOK.

Arsac, G. (1987). L'origine de la démonstration: Essai d'épistémologie didactique. *Recherches en Didactique des Mathématiques, 8*(3), 267-312.

Artigue, M. (1986). The notion of differential for undergraduate students in science. In *Proceedings of the 10th International Conference for the Psychology of Mathematics Education* (Vol. 1, pp. 229-234). London: University of London, Institute of Education.

Artigue, M. (1987). Ingénierie didactique à propos d'équations différentielles. In J. C. Bergeron, N. Herscovics, & C. Kieran (Eds.), *Proceedings of the 11th International Conference for the Psychology of Mathematics Education* (Vol. 3, pp. 236-243). Montréal, Canada: Université de Montréal.

Artigue, M., & Viennot, L. (1987). Students' conceptions and difficulties about differentials. In J. Novak (Ed.), *Proceedings of the Second International Seminar on Misconceptions and Educational Strategies in Science and Mathematics* (Vol. 3, pp. 1-11). Ithaca, NY: Cornell University.

Bachor, D. (1987). Towards a taxonomy of word problems. In J. C. Bergeron, N. Herscovics, & C. Kieran (Eds.), *Proceedings of the 11th International Conference for the Psychology of Mathematics Education* (Vol. 2, pp. 163-169). Montréal, Canada: Université de Montréal.

Balacheff, N. (1982). Preuves et démonstrations en mathématiques au collège. *Recherches en Didactique des Mathématiques, 3*(3), 261-304.

Balacheff, N. (1985). Experimental study of pupils' treatment of refutations in a geometrical context. In L. Streefland (Ed.), *Proceedings of the Ninth International Conference for the Psychology of Mathematics Education* (Vol. 1, pp. 223-229). Utrecht, The Netherlands: State University of Utrecht, Subfaculty of Mathematics, OW & OC.

Balacheff, N. (1987a). Les définitions comme outils dans la résolution de problème. In J. C. Bergeron, N. Herscovics, & C. Kieran (Eds.), *Proceedings of the 11th International Conference for the Psychology of Mathematics Education* (Vol. 3, pp. 187-193). Montréal, Canada: Université de Montréal.

Balacheff, N. (1987b, April). *Towards a problématique for research in mathematics teaching.* Paper presented at the meeting of the Special Interest Group for Research in Mathematics Education, Anaheim, CA.

Balacheff, N. (1988a). Aspects of proof in pupils' practice of school mathematics. In D. Pimm (Ed.), *Mathematics, teachers and children* (pp. 216-230). London: Hodder & Stoughton.

Balacheff, N. (1988b). *Une étude des processus de preuve en mathématique chez les élèves de collège.* Thèse d'état, Université Joseph Fourier, Grenoble, France.

Balacheff, N., & Laborde, C. (1988). Social interactions for experimental studies of pupils' conception: Its relevance for research in didactics of mathematics. In H. G. Steiner & A. Vermandel (Eds.), *Proceedings of the Second Conference on the Theory of Mathematics Education* (pp. 189-195). Bielefeld & Antwerpen: Universität Bielefeld & University of Antwerpen.

Baroody, A. J. (1987). The development of counting strategies for single-digit addition. *Journal for Research in Mathematics Education, 18,* 141-157.

Bauersfeld, H. (1980). Hidden dimensions in the so-called reality of a mathematics classroom. *Educational Studies in Mathematics, 11,* 23-41.

Bauersfeld, H. (1988). Interaction, construction, and knowledge: Alternative perspectives for mathematics education. In D. A. Grouws, T. Cooney, & D. Jones (Eds.), *Research agenda for mathematics education: Vol. 1. Perspectives on research on effective mathematics teaching* (pp. 27-46). Hillsdale, NJ: Erlbaum.

Baulac, Y., Bellemain, F., & Laborde, J. M. (1988). *Cabri Géomètre.* Paris: Cedic-Nathan.

Beaudichon, J. (1982). *La communication sociale chez l'enfant.* Paris: Presses Universitaires de France.

Begle, E. G. (1969). The role of research in the improvement of mathematics education. *Educational Studies in Mathematics, 2,* 232-244.

Bell, A. W. (1976a). A study of pupils' proof-explanations in mathematical situations. *Educational Studies in Mathematics, 7,* 23-40.

Bell, A. W. (1976b). *The learning of general mathematical strategies.* Unpublished manuscript, University of Nottingham.

Bell, A. W., O'Brien, D., & Shiu, C. (1980). Designing teaching in the light of research on understanding. In R. Karplus (Ed.), *Proceedings of the Fourth International Conference for the Psychology of Mathematics Education* (pp. 119-125). Berkeley: University of California.

Ben-Chaim, D., Lappan, G., & Houang, R. T. (1985). Visualizing rectangular solids made of small cubes: Analyzing and effecting students' performance. *Educational Studies in Mathematics, 16,* 389-409.

Ben-Chaim, D., Lappan, G., & Houang, R. T. (1988). The effect of instruction on spatial visualization skills of middle school boys and girls. *American Educational Research Journal, 25,* 51-57.

Ben-Chaim, D., Lappan, G., & Houang, R. T. (1989). Adolescents' ability to communicate spatial information: Analyzing and affecting students' performance. *Educational Studies in Mathematics, 20,* 121-146.

Benacerraf, P., & Putnam, H. (Eds.). (1964). *Philosophy of mathematics: Selected readings.* Englewood Cliffs, NJ: Prentice-Hall.

Bergeron, A., Herscovics, N., & Bergeron, J. C. (1986). Counting tasks involving some hidden elements. In G. Lappan & R. Even (Eds.), *Proceedings of the Eighth Annual Meeting of the North American Branch of the International Group for the Psychology of Mathematics Education* (pp. 21-27). East Lansing: Michigan State University.

Bergeron, J. C., & Herscovics, N. (1988). The kindergartners' understanding of discrete quantity. In A. Borbàs (Ed.), *Proceedings of the 12th International Conference for the Psychology of Mathematics Education* (Vol. 1, pp. 162-169). Veszprém, Hungary: Ferenc Genzwein, OOK.

Bergeron, J. C., Herscovics, N., & Bergeron, A. (1986). The kindergartner's symbolization of numbers. In G. Lappan & R. Even (Eds.), *Proceedings of the Eighth Annual Meeting of the North American Branch of the International Group for the Psychology of Mathematics Education* (pp. 35-41). East Lansing: Michigan State University.

Bergeron, J. C., Herscovics, N., & Moser, J. (1984). Long term evolution of students' conceptions: An example from addition and subtraction. In M. Carss (Ed.), *Proceedings of the Fifth International Congress on Mathematical Education* (pp. 275-277). Boston: Birkhäuser.

Bessot, A., & Eberhard, M. (1986). Adaptation de la perspective à une situation complexe par des élèves de 9-12 ans. *European Journal of Psychology of Education, 1*(2), 83-96.

Bishop, A. J. (1978). Visualising and mathematics in a pre-technological culture. In E. Cohors-Fresenborg & I. Wachsmuth (Eds.), *Proceedings of the Second International Conference for the Psychology of Mathematics Education* (pp. 69-80). Osnabrück, FRG: Osnabrück Schriften zur Mathematik.

Bishop, A. J. (1979). Visual abilities and mathematics learning. In D. Tall (Ed.), *Proceedings of the Third International Conference for the Psychology of Mathematics Education* (pp. 21-26). Warwick, England: Warwick University, Mathematics Education Research Centre.

Bishop, A. J. (1980). Spatial abilities and mathematics education--A review. *Educational Studies in Mathematics, 11*, 257-269.

Bishop, A. J. (1983). Space and geometry. In R. Lesh & M. Landau (Eds.), *Acquisition of mathematics concepts and processes* (pp. 125-203). New York: Academic Press.

Bishop, A. J. (1985). The social psychology of mathematics education. In L. Streefland (Ed.), *Proceedings of the Ninth International Conference for the Psychology of Mathematics Education* (Vol. 2, pp. 1-13). Utrecht, The Netherlands: State University of Utrecht, Subfaculty of Mathematics, OW & OC.

Bishop, A. J. (1989). A review of research on visualization in mathematics education. *Focus on Learning Problems in Mathematics, 11*(1 & 2), 7-16.

Booth, L. R. (1981). Strategies and errors in generalised arithmetic. In C. Comiti & G. Vergnaud (Eds.), *Proceedings of the Fifth International Conference for the Psychology of Mathematics Education* (pp. 140-146). Grenoble, France: Institut IMAG.

Booth, L. R. (1982). Developing a teaching module in beginning algebra. In A. Vermandel (Ed.), *Proceedings of the Sixth International Conference for the Psychology of Mathematics Education* (pp. 280-285). Antwerp, Belgium: Universitaire Instelling Antwerpen.

Booth, L. R. (1983). A diagnostic teaching programme in elementary algebra: Results and implications. In R. Hershkowitz (Ed.), *Proceedings of the Seventh International Conference for the Psychology of Mathematics Education* (pp. 307-312). Rehovot, Israel: Weizmann Institute of Science.

Booth, L. R. (1984). *Algebra: Children's strategies and errors.* Windsor, UK: NFER-Nelson.

Braconne, A., & Dionne, J. J. (1987). Secondary school students' and teachers' understanding of demonstration in geometry. In J. C. Bergeron, N. Herscovics, & C. Kieran (Eds.), *Proceedings of the 11th International Conference for the Psychology of Mathematics Education* (Vol. 3, pp. 109-116). Montréal, Canada: Université de Montréal.

Brousseau, G. (1981). Problèmes de l'enseignement des décimaux. *Recherches en Didactique des Mathématiques, 2*(1), 37-127.

Brousseau, G. (1984, November). *The crucial role of the didactical contract in the analysis and construction of situations in teaching and learning mathematics* (Occasional Paper No. 54). Bielefeld, FRG: Universität Bielefeld, Institut für Didaktik der Mathematik.

Brousseau, G. (1986a). Basic theory and methods in the didactics of mathematics. In P. F. L. Verstappen (Ed.), *Second Conference on Systematic Co-operation Between Theory and Practice in Mathematics Education: Part 1. Report* (pp. 109-161). Enschede, The Netherlands: Instituut voor Leerplanontwikkeling.

Brousseau, G. (1986b). Fondements et méthodes de la didactique des mathematiques. *Recherches en Didactique des Mathématiques, 7*(2), 33-115.

Brouwer, L. E. J. (1964). Intuitionism and formalism. In P. Benacerraf & H. Putnam (Eds.), *Philosophy of mathematics: Selected readings* (pp. 77-89). Englewood Cliffs, NJ: Prentice-Hall. (Original work published 1913)

Brown, G. (1982). The spoken language. In R. Carter (Ed.), *Linguistics and the teacher* (pp. 75-87). London: Routledge & Kegan Paul.

Burger, W. F., & Shaughnessy, J. M. (1986). Characterizing the van Hiele levels of development in geometry. *Journal for Research in Mathematics Education, 17,* 31-48.

Burton, L., Cooper, M., & Leder, G. (1986). Representations of three-dimensional figures by mathematics teachers in-training. In *Proceedings of the 10th International Conference for the Psychology of Mathematics Education* (Vol. 1, pp. 81-86). London: University of London, Institute of Education.

Buxton, L. (1981). *Do you panic about maths?* London: Heinemann.

Carnap, R. (1964). The logicist foundations of mathematics. In P. Benacerraf & H. Putnam (Eds.), *Philosophy of mathematics: Selected readings* (pp. 41-52). Englewood Cliffs, NJ: Prentice-Hall. (Original work published 1931)

Carpenter, T. P., Hiebert, J., & Moser, J. M. (1981). Problem structure and first grade children's initial solution process for simple addition and subtraction problems. *Journal for Research in Mathematics Education, 12,* 27-39.

Carpenter, T. P., & Moser, J. M. (1983). The acquisition of addition and subtraction concepts. In R. Lesh & M. Landau (Eds.), *Acquisition of mathematical concepts and processes* (pp. 7-44). New York: Academic Press.

Carpenter, T. P., & Moser, J. M. (1984). The acquisition of addition and subtraction concepts in grades one through three. *Journal for Research in Mathematics Education, 15,* 179-202.

Carpenter, T. P., Moser, J. M., & Romberg, T. A. (Eds.). (1982). *Addition and subtraction: A cognitive perspective.* Hillsdale, NJ: Erlbaum.

Carraher, T. N. (1988). Street mathematics and school mathematics. In A. Borbàs (Ed.), *Proceedings of the 12th International Conference for the Psychology of Mathematics Education* (Vol. 1, pp. 1-23). Veszprém, Hungary: Ferenc Genzwein, OOK.

Carraher, T. N., Carraher, D. W., & Schliemann, A. D. (1987). Written and oral mathematics. *Journal for Research in Mathematics Education, 18,* 83-97.

Chaiklin, S., & Lesgold, S. (1984, April). *Prealgebra students' knowledge of algebraic tasks with arithmetic expressions.* Paper presented at the annual meeting of the American Educational Research Association, New Orleans, LA.

Chalouh, L., & Herscovics, N. (1988). Teaching algebraic expressions in a meaningful way. In A. Coxford (Ed.), *The ideas of algebra, K-12* (1988 Yearbook of the National Council of Teachers of Mathematics, pp. 33-42). Reston, VA: NCTM.

Chevallard, Y. (1985a). *La transposition didactique.* Grenoble, France: La Pensée Sauvage.

Chevallard, Y. (1985b). Le passage de l'arithmétique a l'algébrique dans l'enseignement des mathématiques au collège. *Petit x, 5,* 51-94.

Chevallard, Y., & Conne, F. (1984). Jalons à propos d'algèbre. *Interactions Didactiques, 3,* 1-54 (Universités de Genève et de Neuchâtel).

Clement, J. (1982). Algebra word problem solutions: Thought processes underlying a common misconception. *Journal for Research in Mathematics Education, 13,* 16-30.

Clement, J. (1985). Misconceptions in graphing. In L. Streefland (Ed.), *Proceedings of the Ninth International Conference for the Psychology of Mathematics Education* (Vol. 1, pp. 369-375). Utrecht, The Netherlands: State University of Utrecht, Subfaculty of Mathematics, OW & OC.

Clement, J., Lochhead, J., & Monk, G. (1981). Translation difficulties in learning mathematics. *American Mathematical Monthly, 88,* 286-290.

Cobb, P. (1986). Contexts, goals, beliefs, and learning mathematics. *For the Learning of Mathematics, 6*(2), 2-9.

Cohors-Fresenborg, E. (1978). Learning problem solving by developing automata networks. In E. Cohors-Fresenborg & I. Wachsmuth (Eds.), *Proceedings of the Second International Conference for the Psychology of Mathematics Education* (pp. 85-92). Osnabrück, FRG: Osnabrück Schriften zur Mathematik.

Cohors-Fresenborg, E. (1982). The understanding of algorithmic concepts on the basis of elementary actions. In A. Vermandel (Ed.), *Proceedings of the Sixth International Conference for the Psychology of Mathematics Education* (pp. 2-7). Antwerp, Belgium: Universitaire Instelling Antwerpen.

Collis, K. F. (1974, June). *Cognitive development and mathematics learning.* Paper presented at the Psychology of Mathematics Workshop, Centre for Science Education, Chelsea College, London.

Collis, K. F. (1975). *The development of formal reasoning.* Unpublished manuscript, University of Newcastle, Australia.

Collis, K. (1986). Learning intellectual skills and school mathematics: A psychological viewpoint. In *Proceedings of the 10th International Conference for the Psychology of Mathematics Education* (Vol. 1, pp. 1-23). London: University of London, Institute of Education.

Combettes, B.   (1983).   *Pour une grammaire textuelle: La progression thématique.* Brussels: A. De Boeck & J. Duculot.

Confrey, J.   (1988).   Multiplication and splitting: Their role in understanding exponential functions. In M. J. Behr, C. B. Lacampagne, & M. M. Wheeler (Eds.), *Proceedings of the 10th Annual Meeting of the North American Branch of the International Group for the Psychology of Mathematics Education* (pp. 250-259). DeKalb: Northern Illinois University.

Conroy, J. S.   (1981).   Learning language and mathematical structure in the infants school. *Research in Mathematics Education in Australia, 2,* 203-212.

Cooney, T. J.   (1985).   A beginning teacher's view of problem solving. *Journal for Research in Mathematics Education, 16,* 324-336.

Cooper, M.   (1984).   The mathematical "reversal error" and attempts to correct it. In B. Southwell, R. Eyland, M. Cooper, J. Conroy, & K. Collis (Eds.), *Proceedings of the Eighth International Conference for the Psychology of Mathematics Education* (pp. 162-171). Sydney: Mathematical Association of New South Wales.

Cooper, M., & Sweller, J.   (1989).   Secondary school students' representations of solids. *Journal for Research in Mathematics Education, 20,* 202-212.

Cornu, B.   (1981).   Apprentissage de la notion de limite: Modèles spontanés et modèles propres. In C. Comiti & G. Vergnaud (Eds.), *Proceedings of the Fifth International Conference for the Psychology of Mathematics Education* (pp. 322-329). Grenoble, France: Institut IMAG.

Davis, P. J., & Hersh, R.   (1981).   *The mathematical experience.* Boston: Birkhäuser.

Davis, R. B.   (1975).   Cognitive processes involved in solving simple algebraic equations. *Journal of Children's Mathematical Behavior, 1*(3), 7-35.

Davydov, V. V., & Andronov, V. P.   (1981).   *Psychological conditions of the origination of ideal actions* (Project Paper 81-2; English translation). Madison: University of Wisconsin, Wisconsin Research and Development Center for Individualized Schooling.

De Avila, E. A.   (1988).   Bilingualism, cognitive function, and language minority group membership. In R. R. Cocking & J. P. Mestre (Eds.), *Linguistic and cultural influences on learning mathematics* (pp. 101-122). Hillsdale, NJ: Erlbaum.

De Corte, E., & Verschaffel, L.   (1985).   Beginning first graders' initial representation of arithmetic word problems. *Journal of Mathematical Behavior, 4,* 3-21.

De Corte, E., & Verschaffel, L.   (1987a).   The effect of semantic structure on first graders' strategies for solving addition and subtraction word problems. *Journal for Research in Mathematics Education, 18,* 363-381.

De Corte, E., & Verschaffel, L.   (1987b).   The effects of semantic and non-semantic factors on young children's solutions of elementary addition and subtraction word problems. In J. C. Bergeron, N. Herscovics, & C. Kieran (Eds.), *Proceedings of the 11th International Conference for the Psychology of Mathematics Education* (Vol. 2, pp. 375-381). Montréal, Canada: Université de Montréal.

De Corte, E., & Verschaffel, L. (1987c). Children's problem-solving skills and processes with respect to elementary arithmetic word problems. In E. De Corte, H. Lodewijks, P. Span, & R. Parmentier (Eds.), *Learning and instruction: European research in an international context* (Vol. 1, pp. 300-308). Leuven, Belgium: Leuven University Press.

De Corte, E., Verschaffel, L. & De Win, L. (1985). The influence of rewording verbal problems on childrens' representations and solutions. *Journal of Educational Psychology, 77,* 460-470.

De Villiers, M. D., & Njisane, R. M. (1987). The development of geometric thinking among high school pupils in Kwazulhu. In J. C. Bergeron, N. Herscovics, & C. Kieran (Eds.), *Proceedings of the 11th International Conference for the Psychology of Mathematics Education* (Vol. 3, pp. 117-123). Montréal, Canada: Université de Montréal.

Dodwell, P. C. (1959). The spatial concepts of the child. *Mathematics Teaching, 9,* 5-14.

Douady, R. (1980). Approche des nombres reals en situation d'apprentissage scolaire: Enfants de Gállans. *Recherches en Didactique des Mathématiques, 1*(1), 77-111.

Douady, R. (1985). The interplay between different settings: Tool-object dialectic in the extension of mathematical ability--Examples from elementary school teaching. In L. Streefland (Ed.), *Proceedings of the Ninth International Conference for the Psychology of Mathematics Education* (Vol. 2, pp. 33-52). Utrecht, The Netherlands: State University of Utrecht, Subfaculty of Mathematics, OW & OC.

Douady, R. (1986a). Concerning conceptions of area (pupils aged 9 to 11). In *Proceedings of the 10th International Conference for the Psychology of Mathematics Education* (Vol. 1, pp. 253-258). London: University of London, Institute of Education.

Douady, R. (1986b). Jeux de cadre et dialectique outil-objet. *Recherches en Didactique des Mathématiques, 7*(2), 5-31.

Dreyfus, H. L., & Dreyfus, S. E. (1986). *Mind over machine: The power of human intuition and expertise in the area of the computer.* New York: Free Press.

Dreyfus, T. (1984). How to use a computer to teach mathematical concepts. In J. M. Moser (Ed.), *Proceedings of the Sixth Annual Meeting of the North American Branch of the International Group for the Psychology of Mathematics Education* (pp. 239-246). Madison: University of Wisconsin.

Dreyfus, T. (in press). Justification and proof in adult education. *Zentralblatt für Didaktik der Mathematik.*

Dreyfus, T., & Eisenberg, T. (1981). Function concepts: Intuitive baseline. In C. Comiti & G. Vergnaud (Eds.), *Proceedings of the Fifth International Conference for the Psychology of Mathematics Education* (pp. 183-188). Grenoble, France: Institut IMAG.

Dreyfus, T., & Eisenberg, T. (1982). Intuitive functional concepts: A baseline study on intuitions. *Journal for Research in Mathematics Education, 13,* 360-380.

Dreyfus, T., & Eisenberg, T. (1983). The function concept in college students: Linearity, smoothness and periodicity. *Focus on Learning Problems in Mathematics, 5*(3 & 4), 119-132.

Drouhard, J.-P., Lymberopoulou-Fioravantes, H., Nikolakarou, H., & Paquelier, Y. (1988). Quelques developpements récents des recherches sur la discussion autour de problèmes. In A. Borbàs (Ed.), *Proceedings of the 12th International Conference for the Psychology of Mathematics Education* (Vol. 1, pp. 247-254). Veszprém, Hungary: Ferenc Genzwein, OOK.

Dubinsky, E. (1988). On helping students construct the concept of quantification. In A. Borbàs (Ed.), *Proceedings of the 12th International Conference for the Psychology of Mathematics Education* (Vol. 1, pp. 255-262). Veszprém, Hungary: Ferenc Genzwein, OOK.

Dubinsky, E., & Lewin, P. (1986). Reflective abstraction and mathematics education: The genetic decomposition of induction and compactness. *Journal of Mathematical Behavior, 5*(1), 55-92.

Eisenberg, T. & Dreyfus, T. (Eds.). (1989). Visualization in the mathematics curriculum [Special issue]. *Focus on Learning Problems in Mathematics, 11*(1 & 2).

Ernest, P. (1985). The philosophy of mathematics and mathematics education. *International Journal of Mathematical Education in Science and Technology, 16*(5), 603-612.

Eshkol, N., & Wachman, M. (1973). *Moving, writing and reading.* Tel Aviv: Tel Aviv University Press.

Figueras, O., & Waldegg, G. (1984). A first approach to measuring (children between 11-13 years old). In J. M. Moser (Ed.), *Proceedings of the Sixth Annual Meeting of the North American Branch of the International Group for the Psychology of Mathematics Education* (pp. 95-99). Madison: University of Wisconsin.

Filloy, E., & Rojano, T. (1984). From an arithmetical to an algebraic thought. In J. M. Moser (Ed.), *Proceedings of the Sixth Annual Meeting of the North American Branch of the International Group for the Psychology of Mathematics Education* (pp. 51-56). Madison: University of Wisconsin.

Filloy, E., & Rojano, T. (1985a). Obstructions to the acquisition of elemental algebraic concepts and teaching strategies. In L. Streefland (Ed.), *Proceedings of the Ninth International Conference for the Psychology of Mathematics Education* (Vol. 1, pp. 154-158). Utrecht, The Netherlands: State University of Utrecht, Subfaculty of Mathematics, OW & OC.

Filloy, E., & Rojano, T. (1985b). Operating on the unknown and models of teaching. In S. K. Damarin & M. Shelton (Eds.), *Proceedings of the Seventh Annual Meeting of the North American Branch of the International Group for the Psychology of Mathematics Education* (pp. 75-79). Columbus: Ohio State University.

Fischbein, E. (1987). *Intuition in science and mathematics: An educational approach.* Dordrecht, The Netherlands: Reidel.

Fischbein, E., & Kedem, I. (1982). Proof and certitude in the development of mathematical thinking. In A. Vermandel (Ed.), *Proceedings of the Sixth International Conference for the Psychology of Mathematics Education* (pp. 128-131). Antwerp, Belgium: Universitaire Instelling Antwerpen.

Frazier, L., & Raynier, K. (1982). Making and correcting errors during sentence comprehension. *Cognitive Psychology, 14,* 178-210.

Freudenthal, H. (1971). Geometry between the devil and the deep sea. *Educational Studies in Mathematics, 3,* 413-435.

Freudenthal, H. (1973). *Mathematics as an educational task.* Dordrecht, The Netherlands: Reidel.

Freudenthal, H. (1983). *Didactical phenomenology of mathematical structures.* Dordrecht, The Netherlands: Reidel.

Friedlander, A., Fitzgerald, W. M., & Lappan, G. (1984). The growth of similarity concepts at the sixth grade level. In J. M. Moser (Ed.), *Proceedings of the Sixth Annual Meeting of the North American Branch of the International Group for the Psychology of Mathematics Education* (pp. 127-132). Madison: University of Wisconsin.

Fuson, K. C. (1982). An analysis of the counting-on solution procedure in addition. In T. P. Carpenter, J. M. Moser, & T. A. Romberg (Eds.), *Addition and subtraction: A cognitive perspective* (pp. 67-81). Hillsdale, NJ: Erlbaum.

Fuson, K. C. (1988). *Children's counting and concepts of number.* New York: Springer-Verlag.

Fuson, K. C., Richards, J., & Briars, D. J. (1982). The acquisition and elaboration of the number word sequence. In C. Brainerd (Ed.), *Progress in cognitive development research: Vol. 1. Children's logical and mathematical cognition* (pp. 33-92). New York: Springer-Verlag.

Gallou-Dumiel, E. (1987). Théorème de Thalès et ordinateur. In J. C. Bergeron, N. Herscovics, & C. Kieran (Eds.), *Proceedings of the 11th International Conference for the Psychology of Mathematics Education* (Vol. 2, pp. 10-16). Montréal, Canada: Université de Montréal.

Gardiner, A. (1982). *Infinite processes.* New York: Springer.

Gaulin, C. (1985). The need for emphasizing various graphical representations of 3-dimensional shapes and relations. In L. Streefland (Ed.), *Proceedings of the Ninth International Conference for the Psychology of Mathematics Education* (Vol. 2, pp. 53-71). Utrecht, The Netherlands: State University of Utrecht, Subfaculty of Mathematics, OW & OC.

Gelman, R. (1972). The nature and development of early number concepts. In H. Reese (Ed.), *Advances in child development and behavior* (Vol. 7, pp. 115-167). New York: Academic Press.

Gelman, R. (1977). How young children reason about small numbers. In N. J. Castellan, D. P. Pisoni, & G. R. Potts (Eds.), *Cognitive theory* (Vol. 2, pp. 219-238). Hillsdale, NJ: Erlbaum.

Gelman, R. (1982). Basic numerical abilities. In R. J. Sternberg (Ed.), *Advances in psychology of human intelligence* (Vol. 1, pp. 181-205). Hillsdale, NJ: Erlbaum.

Gelman, R., & Gallistel, C. R. (1978). *The child's understanding of number.* Cambridge, MA: Harvard University Press.

Ginsburg, H. (1975). Young children's informal knowledge of mathematics. *Journal of Children's Mathematical Behavior, 1*(3), 63-156.

Ginsburg, H. (1976). Learning difficulties in children's arithmetic: A clinical cognitive approach. In A. O. Osborne (Ed.), *Models for learning mathematics.* Columbus, OH: SMEAC/ERIC.

Ginsburg, H. (1977). The psychology of arithmetic thinking. *Journal of Children's Mathematical Behavior, 1*(4), 1-89.

Ginsburg, H. (1982). *Children's arithmetic: The learning process.* Austin, TX: Pro-Ed.

Glaeser, G. (1981). Epistémologie des nombres relatifs. *Recherches en Didactique des Mathématiques, 2*(3), 303-346.

Goddijn, A., & Kindt, M. (1985). Space geometry doesn't fit in the book. In L. Streefland (Ed.), *Proceedings of the Ninth International Conference for the Psychology of Mathematics Education* (Vol. 1, pp. 171-182). Utrecht, The Netherlands: State University of Utrecht, Subfaculty of Mathematics, OW & OC.

Gödel, K. (1964). What is Cantor's continuum problem? In P. Benacerraf & H. Putnam (Eds.), *Philosophy of mathematics: Selected readings* (pp. 470-485). Englewood Cliffs, NJ: Prentice-Hall. (Original work published 1947)

Gréco, P. (1962). Quantité et quotité. In P. Gréco & A. Morf (Eds.), *Structures numériques élémentaires* (pp. 1-70). Paris: Presses Universitaires de France.

Greeno, J. G. (1982, March). *A cognitive learning analysis of algebra.* Paper presented at the annual meeting of the American Educational Research Association, Boston, MA.

Grenier, D. (1985). Middle school pupils' conceptions about reflections according to a task of construction. In L. Streefland (Ed.), *Proceedings of the Ninth International Conference for the Psychology of Mathematics Education* (Vol. 1, pp. 183-188). Utrecht, The Netherlands: State University of Utrecht, Subfaculty of Mathematics, OW & OC.

Grenier, D. (1987, July). The pupils' conceptions on axial symmetry: An individual activity in a classroom. In R. Hershkowitz & S. Vinner (Eds.), *Geometry working group report from the 10th conference and some subsequent reactions* (pp. 14-27). Report presented at the 11th International Conference for the Psychology of Mathematics Education, Montréal, Canada.

Grenier, D., & Laborde, C. (1988). Transformations géométriques: Le cas de la symétrie orthogonale. In G. Vergnaud, G. Brousseau, & M. Hulin (Eds.), *Actes du*

Colloque du GRECO Didactique et Acquisition des Connaissances Scientifiques (pp. 65-86). Grenoble, France: La Pensée Sauvage.

Groen, G. J., & Parkman, J. M. (1972). A chronometric analysis of simple addition. *Psychological Review, 79,* 329-343.

Guillerault, M., & Laborde, C. (1986). A study of pupils reading geometry. In F. Lowenthal & F. Vandamme (Eds.), *Pragmatics and education* (pp. 223-238). London: Plenum.

Gutierrez, A., & Jaime, A. (1987). Estudio de las characterísticas de los niveles de van Hiele. In J. C. Bergeron, N. Herscovics, & C. Kieran (Eds.), *Proceedings of the 11th International Conference for the Psychology of Mathematics Education* (Vol. 3, pp. 131-137). Montréal, Canada: Université de Montréal.

Hadamard, J. (1949). *An essay on the psychology of invention in the mathematical field.* Princeton, NJ: Princeton University Press.

Hadas, N., Dreyfus, T., & Friedlander, A. (1983). Euclidean geometry for average ability children. In R. Hershkowitz (Ed.), *Proceedings of the Seventh International Conference for the Psychology of Mathematics Education* (pp. 253-258). Rehovot, Israel: Weizmann Institute of Science.

Halliday, M. A. K. (1975). Some aspects of sociolinguistics. In E. Jacobsen (Ed.), *Interactions between linguistics and mathematical education* (UNESCO Report No. ED-74/CONF.808, pp. 64-73). Paris: UNESCO.

Harper, E. (1987). Ghosts of Diophantus. *Educational Studies in Mathematics, 18,* 75-90.

Harris, M. (1987, July). Math in work--Another look at rectangles. In R. Hershkowitz & S. Vinner (Eds.), *Geometry working group report from the 10th conference and some subsequent reactions* (pp. 9-10). Report presented at the 11th International Conference for the Psychology of Mathematics Education, Montréal, Canada.

Hatano, G. (1982). Learning to add and subtract: A Japanese perspective. In T. P. Carpenter, J. M. Moser, & T. A. Romberg (Eds.), *Addition and subtraction: A cognitive perspective* (pp. 211-223). Hillsdale, NJ: Erlbaum.

Hausmann, K. (1985) Iterative and recursive models of thinking in mathematical problem solving processes. In L. Streefland (Ed.), *Proceedings of the Ninth International Conference for the Psychology of Mathematics Education* (Vol. 1, pp. 18-23). Utrecht, The Netherlands: State University of Utrecht, Subfaculty of Mathematics, OW & OC.

Heid, M. K. (1988). *"Algebra with Computers": A description and an evaluation of student performance and attitudes* (Report submitted to the State College Area School District Board of Education). State College: The Pennsylvania State University.

Heid, M. K., Sheets, C., Matras, M. A., & Menasian, J. (1988, April). *Classroom and computer lab interaction in a computer-intensive algebra curriculum.* Paper presented at the annual meeting of the American Educational Research Association, New Orleans, LA.

Herscovics, N., & Bergeron, J. C. (1988a). An extended model of understanding. In M. J. Behr, C. B. Lacampagne, & M. M. Wheeler (Eds.), *Proceedings of the 10th Annual Meeting of the North American Branch of the International Group for the Psychology of Mathematics Education* (pp. 15-22). DeKalb: Northern Illinois University.

Herscovics, N., & Bergeron, J. C. (1988b). The kindergartner's understanding of the notion of rank. In A. Borbàs (Ed.), *Proceedings of the 12th International Conference for the Psychology of Mathematics Education* (Vol. 1, pp. 385-392). Veszprém, Hungary: Ferenc Genzwein, OOK.

Herscovics, N., & Bergeron, J. C. (1989). The kindergartner's understanding of cardinal number: An international study. In G. Vergnaud, J. Rogalski, & M. Artigue (Eds.), *Proceedings of the 13th Annual Meeting of the International Group for the Psychology of Mathematics Education* (Vol. 2, pp. 91-98). Paris: Laboratoire PSYDEE.

Herscovics, N., Bergeron, J. C., & Bergeron, A. (1986). The kindergartner's perception of the invariance of number under various transformations. In G. Lappan & R. Even (Eds.), *Proceedings of the Eighth Annual Meeting of the North American Branch of the International Group for the Psychology of Mathematics Education* (pp. 28-34). East Lansing: Michigan State University.

Herscovics, N., & Kieran, C. (1980). Constructing meaning for the concept of equation. *Mathematics Teacher, 73,* 572-580.

Hershkowitz, R. (1987). The acquisition of concepts and misconceptions in basic geometry--Or when "a little learning is a dangerous thing." In J. D. Novak (Ed.), *Proceedings of the Second International Seminar on Misconceptions and Educational Strategies in Science and Mathematics* (Vol. 3, pp. 238-251). Ithaca, NY: Cornell University.

Hershkowitz, R. (1989). Visualization in geometry: Two sides of the coin. *Focus on Learning Problems in Mathematics, 11*(1 & 2), 61-75.

Hershkowitz, R., & Vinner, S. (1983). The role of critical and non-critical attributes in the concept-image of geometrical concepts. In R. Hershkowitz (Ed.), *Proceedings of the Seventh International Conference for the Psychology of Mathematics Education* (pp. 223-228). Rehovot, Israel: Weizmann Institute of Science.

Hershkowitz, R., & Vinner, S. (Eds.). (1987, July). *Geometry working group report from the 10th conference and some subsequent reactions.* Report presented at the 11th International Conference for the Psychology of Mathematics Education, Montréal, Canada.

Hershkowitz, R., Vinner, S. & Bruckheimer, M. (1987). Activities with teachers based on cognitive research. In M. M. Lindquist & A. P. Shulte (Eds.), *Learning and teaching geometry K-12* (1987 Yearbook of the National Council of Teachers of Mathematics, pp. 222-235). Reston, VA: NCTM.

Heyting, A. (1956). *Intuitionism: An introduction.* Amsterdam: North Holland.

Hillel, J. (1986). Procedural thinking by children aged 8-12, using turtle-geometry. In *Proceedings of the 10th International Conference for the Psychology of Mathematics Education* (Vol. 1, pp. 433-438). London: University of London, Institute of Education.

Hillel, J., & Kieran, C. (1987). Schemas used by 12-year-olds solving selected turtle-geometry tasks. *Recherches en Didactique des Mathématiques, 8*(1.2), 61-103.

Hilton, P. (1976). Education in mathematics and science today: The spread of false dichotomies. In H. Athen & H. Kunle (Eds.), *Proceedings of the Third International Congress on Mathematical Education* (pp. 75-97). Karlsruhe, FRG: University of Karlsruhe.

Hirabayashi, I., & Shigematsu, K. (1986). Meta-cognition: The role of the "inner teacher." In *Proceedings of the 10th International Conference for the Psychology of Mathematics Education* (Vol. 1, pp. 165-170). London: University of London, Institute of Education.

Hirabayashi, I., & Shigematsu, K. (1987). Meta-cognition: The role of the "inner teacher" (2). In J. C. Bergeron, N. Herscovics, & C. Kieran (Eds.), *Proceedings of the 11th International Conference for the Psychology of Mathematics Education* (Vol. 2, pp. 243-249). Montréal, Canada: Université de Montréal.

Hoffer, A. (1981). Geometry is more than proof. *Mathematics Teacher, 74,* 11-18.

Hoffer, A. (1983). Van Hiele based research. In R. Lesh & M. Landau (Eds.), *Acquisition of mathematics concepts and processes* (pp. 205-227). New York: Academic Press.

Howson, A. G. (Ed.). (1973). *Developments in mathematical education: Proceedings of the Second International Congress on Mathematical Education.* Cambridge: Cambridge University Press.

Howson, A. G. (1984). Seventy-five years of the International Commission on Mathematical Instruction. *Educational Studies in Mathematics, 15,* 75-93.

Hoyles, C. (1985). What is the point of group discussion in mathematics? *Educational Studies in Mathematics, 16,* 205-214.

Hoyles, C. (1987). Geometry and the computer environment. In J. C. Bergeron, N. Herscovics, & C. Kieran (Eds.), *Proceedings of the 11th International Conference for the Psychology of Mathematics Education* (Vol. 2, pp. 60-66). Montréal, Canada: Université de Montréal.

Hoyles, C., & Noss, R. (1987a). Children working in a structured Logo environment: From doing to understanding. *Recherches en Didactique des Mathématiques, 8*(1-2), 131-174.

Hoyles, C., & Noss, R. (1987b). Seeing what matters: Developing an understanding of the concept of parallelogram through a Logo microworld. In J. C. Bergeron, N.

Herscovics, & C. Kieran (Eds.), *Proceedings of the 11th International Conference for the Psychology of Mathematics Education* (Vol. 2, pp. 17-23). Montréal, Canada: Université de Montréal.

Hoyles, C., Noss, R., & Sutherland, R. (1989). A Logo-based microworld for ratio and proportion. In G. Vergnaud, J. Rogalski, & M. Artigue (Eds.), *Proceedings of the 13th International Conference for the Psychology of Mathematics Education* (Vol. 2, pp. 115-122). Paris: Laboratoire PSYDEE.

Hoyles, C., & Sutherland, R. (1986). Peer interaction in a programming environment. In *Proceedings of the 10th International Conference for the Psychology of Mathematics Education* (Vol. 1, pp. 354-359). London: University of London, Institute of Education.

Hoyles, C., & Sutherland, R. (1989). *Logo mathematics in the classroom.* London: Routledge & Kegan Paul.

Hoyles, C., Sutherland, R., & Evans, J. (1985). *The Logo Maths Project: A preliminary investigation of the pupil-centred approach to the learning of Logo in the secondary mathematics classroom, 1983-4.* London: University of London, Institute of Education.

Hoz, R. (1979). The effects of rigidity on the solution of geometrical "problems to prove." In D. Tall (Ed.), *Proceedings of the Third International Conference for the Psychology of Mathematics Education* (pp. 112-117). Warwick, England: Warwick University, Mathematics Education Research Centre.

Jackson, L., & Coutts, J. (1987). Measuring behavioral success avoidance in mathematics dyadic settings. In J. C. Bergeron, N. Herscovics, & C. Kieran (Eds.), *Proceedings of the 11th International Conference for the Psychology of Mathematics Education* (Vol. 1, pp. 84-91). Montréal, Canada: Université de Montréal.

Janvier, C. (1981). Difficulties related to the concept of variable presented graphically. In C. Comiti & G. Vergnaud (Eds.), *Proceedings of the Fifth International Conference for the Psychology of Mathematics Education* (Vol. 1, pp. 189-192). Grenoble, France: Institut IMAG.

Janvier, C. (1982). Approaches to the notion of function in relation to set theory. In G. van Barneveld & H. Krabbendam (Eds.), *Proceedings of the Conference on Functions* (pp. 114-124). Enschede, The Netherlands: National Institute for Curriculum Development.

Janvier, C. (Ed.). (1987). *Problems of representation in the teaching and learning of mathematics.* Hillsdale, NJ: Erlbaum.

Joffe, L., & Foxman, D. (1986). Factors affecting small group performance in problem solving. In *Proceedings of the 10th International Conference for the Psychology of Mathematics Education* (Vol. 1, pp. 360-362). London: University of London, Institute of Education.

Kaput, J. (1987). PME-XI algebra papers: A representational framework. In J. C. Bergeron, N. Herscovics, & C. Kieran (Eds.), *Proceedings of the 11th International*

Conference for the Psychology of Mathematics Education (Vol. 1, pp. 345-354). Montréal, Canada: Université de Montréal.

Kaput, J., Luke, C., Poholsky, J., & Sayer, A. (1987). Multiple representations and reasoning with discrete intensive quantities in a computer-based environment. In J. C. Bergeron, N. Herscovics, & C. Kieran (Eds.), *Proceedings of the 11th International Conference for the Psychology of Mathematics Education* (Vol. 2, pp. 289-295). Montréal, Canada: Université de Montréal.

Karplus, R. (1979). Continuous functions: Students' viewpoints. *European Journal of Science Education, 1,* 379-415.

Kerslake, D. (1977). The understanding of graphs. *Mathematics in School, 6*(2), 22-25.

Kesler, R. (1985). *Teachers' instructional behaviors related to their conceptions of teaching and mathematics and their level of dogmatism: Four case studies.* Unpublished doctoral dissertation, University of Georgia, Athens.

Kieran, C. (1981). Concepts associated with the equality symbol. *Educational Studies in Mathematics, 12,* 317-326.

Kieran, C. (1983). Relationships between novices' views of algebraic letters and their use of symmetric and asymmetric equation-solving procedures. In J. C. Bergeron & N. Herscovics (Ed.), *Proceedings of the Fifth Annual Meeting of the North American Branch of the International Group for the Psychology of Mathematics Education* (Vol. 1, pp. 161-168). Montréal, Canada: Université de Montréal.

Kieran, C. (1984). A comparison between novice and more-expert algebra students on tasks dealing with the equivalence of equations. InJ. M. Moser (Ed.), *Proceedings of the Sixth Annual Meeting of the North American Branch of the International Group for the Psychology of Mathematics Education* (pp. 83-91). Madison: University of Wisconsin.

Kieran, C. (1985). Use of substitution procedure in learning algebraic equation-solving. In S. K. Damarin & M. Shelton (Eds.), *Proceedings of the Seventh Annual Meeting of the North American Branch of the International Group for the Psychology of Mathematics Education* (pp. 145-152). Columbus: Ohio State University.

Kieran, C. (1988). Two different approaches among algebra learners. In A. F. Coxford (Ed.), *The ideas of algebra, K-12* (1988 Yearbook of the National Council of Teachers of Mathematics, pp. 91-96). Reston, VA: NCTM.

Kieran, C. (1989). The early learning of algebra: A structural perspective. In S. Wagner & C. Kieran (Eds.), *Research agenda for mathematics education: Vol. 4. Research issues in the learning and teaching of algebra* (pp. 33-56). Hillsdale, NJ: Erlbaum.

Kieran, C., Hillel, J., & Erlwanger, S. (1986). Perceptual and analytical schemas in solving structural turtle-geometry tasks. In C. Hoyles, R. Noss, & R. Sutherland (Eds.), *Proceedings of the Second International Conference for Logo and Mathematics Education* (pp. 154-161). London: University of London, Institute of Education.

Kilpatrick, J. (1987). What constructivism might be in mathematics education. In J. C. Bergeron, N. Herscovics, & C. Kieran (Eds.), *Proceedings of the 11th International Conference for the Psychology of Mathematics Education* (Vol. 1, pp. 3-27). Montréal, Canada: Université de Montréal.

Kilpatrick, J., & Wirszup, I. (Eds.). (1969). *Soviet studies in the psychology of learning and teaching mathematics* (Vols. 1-6). Stanford, CA: School Mathematics Study Group.

Kintsch, W., & van Dijk, T. (1978). Text comprehension and production. *Psychological Review, 85,* 363-394.

Kirschner, D. (1987). The myth about binary representation in algebra. In J. C. Bergeron, N. Herscovics, & C. Kieran (Eds.), *Proceedings of the 11th International Conference for the Psychology of Mathematics Education* (Vol. 1, pp. 308-315). Montréal, Canada: Université de Montréal.

Klahr, D., & Wallace, J. G. (1976). *Cognitive development: An information processing view.* Hillsdale, NJ: Erlbaum.

Kleiner, I. (1989). Evolution of the function concept: A brief survey. *College Mathematics Journal, 20,* 282-300.

Kramer, E., Hadas, N., & Hershkowitz, R. (1986). Geometrical constructions and the micro-computer. In *Proceedings of the 10th International Conference for the Psychology of Mathematics Education* (Vol. 1, pp. 105-110). London: University of London, Institute of Education.

Küchemann, D. (1981). Algebra. In K. Hart (Ed.), *Children's understanding of mathematics: 11-16* (pp. 102-119). London: John Murray.

Kuhn, T. S. (1970). *The structure of scientific revolutions* (2nd ed.). Chicago: University of Chicago Press.

Laborde, C. (1982). *Langue naturelle et écriture symbolique: Deux codes en interaction dans l'enseignement mathématique.* Thèse d'état, Université Joseph Fourier, Institut IMAG, Grenoble, France.

Laborde, C. (1987). Lecture de textes mathématiques par des élèves (14-15 ans): Une expérimentation. In J. C. Bergeron, N. Herscovics, & C. Kieran (Eds.), *Proceedings of the 11th International Conference for the Psychology of Mathematics Education* (Vol. 3, pp. 194-200). Montréal, Canada: Université de Montréal.

Laborde, C. (1988). Divers aspects de la dimension sociale dans les recherches en didactique des mathématiques. In C. Laborde (Ed.), *Actes du Premier Colloque Franco-Allemand de Didactique des Mathématiques et de l'Informatique* (pp. 67-80). Grenoble: La Pensée Sauvage.

Laborde, C., & Mejias, B. (1985). The construction process of an iteration by middle-school pupils: An experimental approach. In L. Streefland (Ed.), *Proceedings of the Ninth International Conference for the Psychology of Mathematics Education* (Vol. 1, pp. 40-46). Utrecht, The Netherlands: State University of Utrecht, Subfaculty of Mathematics, OW & OC.

Lakatos, I. (1976). *Proofs and refutations.* Cambridge: Cambridge University Press.

Lambdin Kroll, D. (1988). Cooperative mathematical problem solving and metacognition: A case study of three pairs of college aged women. In M. J. Behr, C. B. Lacampagne, & M. M. Wheeler (Eds.), *Proceedings of the 10th Annual Meeting of the North American Branch of the International Group for the Psychology of Mathematics Education* (pp. 229-235). DeKalb: Northern Illinois University.

Leder, G. (1980). Bright girls, mathematics and fear of success. *Educational Studies in Mathematics, 11,* 411-422.

Lee, L. (1987). The status and understanding of generalised algebraic statements by high school students. In J. C. Bergeron, N. Herscovics, & C. Kieran (Eds.), *Proceedings of the 11th International Conference for the Psychology of Mathematics Education* (Vol. 1, pp. 316-323). Montréal, Canada: Université de Montréal.

Lee, L., & Wheeler, D. (1986). High school students' conception of justification in algebra. In G. Lappan & R. Even (Eds.), *Proceedings of the Eighth Annual Meeting of the North American Branch of the International Group for the Psychology of Mathematics Education* (pp. 94-101). East Lansing: Michigan State University.

Lee, L., & Wheeler, D. (1987). *Algebraic thinking in high school students: Their conceptions of generalisation and justification.* Unpublished manuscript, Concordia University, Montréal, Canada.

Lerman, S. (1987). *Investigations, where to now? Or problem-posing and the nature of mathematics* (Perspectives No. 33). Unpublished manuscript, University of Exeter.

Leron, U. (1983a). Some problems in children's Logo learning. In R. Hershkowitz (Ed.), *Proceedings of the Seventh International Conference for the Psychology of Mathematics Education* (pp. 346-351). Rehovot, Israel: Weizmann Institute of Science.

Leron, U. (1983b). Structuring mathematical proofs. *American Mathematical Monthly, 90,* 174-185.

Leron, U. (1985). Heuristic presentations: The role of structuring. *For the Learning of Mathematics, 5*(3), 7-13.

Lesh, R. (1985). Conceptual analysis of mathematical ideas and problem solving processes. In L. Streefland (Ed.), *Proceedings of the Ninth International Conference for the Psychology of Mathematics Education* (Vol. 2, pp. 73-96). Utrecht, The Netherlands: State University of Utrecht, Subfaculty of Mathematics, OW & OC.

Lewis, C. (1980, April). *Kinds of knowledge in algebra.* Paper presented at the annual meeting of the American Education Research Association, Boston.

Lovell, K. (1959). A follow-up study of some aspects of the work of Piaget and Inhelder on child's conception of space. *British Journal of Educational Psychology, 29,* 104-117.

Lowenthal, F., & Vandeputte, C. (1989). Manipulations of Cartesian graphs: A first introduction to analysis. *Focus on Learning Problems in Mathematics, 11*(1 & 2), 89-98.

Ludwig, S. R., & Kieren, T. E. (1985). Logo geometry: Ego syntonic? In S. K. Damarin & M. Shelton (Eds.), *Proceedings of the Seventh Annual Meeting of the North American Branch of the International Group for the Psychology of Mathematics Education* (pp. 345-350). Columbus: Ohio State University.

Maher, C. A., & Beattys, C. B. (1986). Examining the construction of area and its measurement by 10 to 14 year old children. In G. Lappan & R. Even (Eds.), *Proceedings of the Eighth Annual Meeting of the North American Branch of the International Group for the Psychology of Mathematics Education* (pp. 163-168). East Lansing: Michigan State University.

Markovits, Z., Eylon, B., & Bruckheimer, M. (1986). Functions today and yesterday. *For the Learning of Mathematics, 6*(2), 18-24.

Martin, J. L. (1976a). An analysis of some of Piaget's topological tasks from a mathematical point of view. *Journal for Research in Mathematics Education, 7,* 8-24.

Martin, J. L. (1976b). A test with selected topological properties of Piaget's hypothesis concerning the spatial representation of the young child. *Journal for Research in Mathematics Education, 7,* 26-38.

Martin, W. G., & Harel, G. (1989). Proof frames of preservice elementary teachers. *Journal for Research in Mathematics Education, 20,* 41-51.

Mason, J., & Pimm, D. (1984). Generic examples: Seeing the general in the particular. *Educational Studies in Mathematics, 15,* 177-289.

Matos, J. M. (1985). Geometric concepts of Portuguese preservice primary teachers. In L. Streefland (Ed.), *Proceedings of the Ninth International Conference for the Psychology of Mathematics Education* (Vol. 1, pp. 189-194). Utrecht, The Netherlands: State University of Utrecht, Subfaculty of Mathematics, OW & OC.

Mayberry, J. W. (1983). The van Hiele levels of geometric thought in undergraduate preservice teachers. *Journal for Research in Mathematics Education, 14,* 58-69.

Mayer, R. E. (1977). *Thinking and problem solving: An introduction to human cognition and learning.* Glenview, IL: Scott, Foresman.

McGalliard, W. (1983). Selected factors in the conceptual systems of geometry teachers: Four case studies (Doctoral dissertation, University of Georgia, 1982). *Dissertation Abstracts International, 44,* 1364A.

McKnight, C. C., Crosswhite, F. J., Dossey, J. A., Kifer, E., Swafford, J. O, Travers, K. J., & Cooney, T. J. (1987). *The underachieving curriculum: Assessing U.S. school mathematics from an international perspective.* Champaign, IL: Stipes.

Menchinskaya, N. A. (1969). The psychology of mastering concepts. In J. Kilpatrick & I. Wirszup (Eds.), *Soviet studies in the psychology of learning and teaching mathematics: Vol. 1. The learning of mathematical concepts* (pp. 75-92). Stanford, CA: School Mathematics Study Group.

Mestre, J. P. (1988). The role of language comprehension in mathematics and problem solving. In R. R. Cocking & J. P. Mestre (Eds.), *Linguistic and cultural influences on learning mathematics* (pp. 201-220). Hillsdale, NJ: Erlbaum.

Metzler, J., & Shepard, R. N. (1974). Transformational studies of the internal representation of three-dimensional objects. In R. L. Solso (Ed.), *Theories in cognitive psychology: The Loyola Symposium* (pp. 147-201). Potomac, MD: Erlbaum.

Mevarech, Z. R., & Yitschak, D. (1983). Students' misconceptions of the equivalence relationship. In R. Hershkowitz (Ed.), *Proceedings of the Seventh International Conference for the Psychology of Mathematics Education* (pp. 313-318). Rehovot, Israel: Weizmann Institute of Science.

Meyerson, L. N. (1978). Conception of knowledge in mathematics: Interaction with and applications to a teaching methods course (Doctoral dissertation, State University of New York, Buffalo, 1977). *Dissertation Abstracts International, 38,* 733A.

Michener, E. R. (1978). *Understanding understanding mathematics* (A.I. Memo No. 488; Logo Memo No. 50). Boston, MA: Massachusetts Institute of Technology, Artificial Intelligence Laboratory.

Mitchelmore, M. C. (1980a). Three dimensional geometrical drawing in three cultures. *Educational Studies in Mathematics, 11,* 205-216.

Mitchelmore, M. C. (1980b). Representation of regular space figures. *Journal for Research in Mathematics Education, 11,* 83-93.

Mitchelmore, M. C. (1983). 3D drawings and the parallel concept. In R. Hershkowitz (Ed.), *Proceedings of the Seventh International Conference for the Psychology of Mathematics Education* (pp. 40-44). Rehovot, Israel: Weizmann Institute of Science.

Morange, D. (1985). Quelques aspects des fonctions du langage dans la résolution de problèmes additifs. *Séminaire de Didactique des Mathématiques et de l'Informatique* (LSD-IMAG, Institut Fourier, Grenoble, France, année 1985-1986), pp. 3-38.

Movshovitz-Hadar, N. (1988). Stimulating presentations of theorems followed by responsive proofs. *For the Learning of Mathematics, 8*(2), 12-19

Mukhopadhyay, S. (1987, July). *On the drawing of solid stimuli: The scaling of responses of rural Indians from specific occupational backgrounds.* Paper presented as a poster at the 11th International Conference for the Psychology of Mathematics Education, Montréal, Canada.

Nesher, P. (1982). Levels of description in the analysis of addition and subtraction. In T. P. Carpenter, J. M. Moser, & T. A. Romberg (Eds.), *Addition and subtraction: A cognitive perspective.* Hillsdale, NJ: Erlbaum.

Nesher, P., & Katriel, T. (1977). A semantic analysis of addition and subtraction word problems in arithmetic. *Educational Studies in Mathematics, 8,* 251-269.

Nesher, P., & Katriel, T. (1978). Two cognitive modes in arithmetic word problem solving. In E. Cohors-Fresenborg & I. Wachsmuth (Eds.), *Proceedings of the Second*

International Conference for the Psychology of Mathematics Education (pp. 189-204). Osnabrück, FRG: Osnabrück Schriften zur Mathematik.

Nesher, P., & Teubal, E. (1975). Verbal cues as an interfering factor in verbal problem solving. *Educational Studies in Mathematics, 6,* 41-51.

Noddings, N. (1985). Small groups as a setting for research on mathematical problem solving. In E. A. Silver (Ed.), *Teaching and learning mathematical problem solving: Multiple research perspectives* (pp. 345-359). Hillsdale, NJ: Erlbaum.

Noss, R. (1987). Children's learning of geometrical concepts through Logo. *Journal for Research in Mathematics Education, 18,* 343-362.

Oléron, P. (1978). *Langage et développement mental.* Brussels: Pierre Mardaga Editeurs.

Orton, R. E. (1988). Two theories of "theory" in mathematics education: Using Kuhn and Lakatos to examine four fundamental issues. *For the Learning of Mathematics, 8*(2), 36-43.

Osta, I. (1987). L'outil informatique et l'enseignement de la géométrie de l'espace. In J. C. Bergeron, N. Herscovics, & C. Kieran (Eds.), *Proceedings of the 11th International Conference for the Psychology of Mathematics Education* (Vol. 2, pp. 31-38). Montréal, Canada: Université de Montréal.

Osta, I. (1988). *L'ordinateur comme outil d'aide à l'enseignement une séquence didactique pour l'enseignement du repérage dans l'espace à l'aide de lodiciels graphiques.* Doctoral dissertation, Université Joseph Fourier, Institut IMAG, Grenoble, France.

Owens, J. (1986). *A study of four preservice secondary mathematics teachers' constructs of mathematics and mathematics teaching.* Unpublished doctoral dissertation. University of Georgia, Athens.

Petitto, A. (1979). The role of formal and non-formal thinking in doing algebra. *Journal of Children's Mathematical Behavior, 2*(2), 69-82.

Piaget, J. (1945). *La formation du symbole.* Neuchâtel: Delachaux et Niestlé.

Piaget, J. (1949). *Introduction à l'épistémologie génétique.* Paris: Presses Universitaires de France.

Piaget, J. (1952). *The child's conception of number.* London: Routledge & Kegan Paul.

Piaget, J. (1970). Piaget's theory. In P. Musson (Ed.), *Carmichael's manual of child psychology* (Vol. 1, pp. 703-732). New York: Wiley.

Piaget, J. (1973). Comments on mathematical education. In A. G. Howson (Ed.), *Developments in mathematical education: Proceedings of the Second International Congress on Mathematical Education* (pp. 79-87). Cambridge: Cambridge University Press.

Piaget, J., & Inhelder, B. (1963). Les opérations intellectuelles et leur développement. In P. Oléron, J. Piaget, B. Inhelder, & P. Gréco (Eds.), *Traité de psychologie expérimentale* (Vol. 7, pp. 109-153). Paris: Presses Universitaires de France.

Piaget, J., & Inhelder, B. (1967). *The child's conception of space.* New York: Norton.

Piaget, J., Inhelder, B., & Szeminska, A. (1960). *The child's conception of geometry.* London: Routledge & Kegan Paul.

Piaget, J., & Szeminska, A. (1941). *La genèse du nombre chez l'enfant.* Neuchâtel: Delachaux et Niestlé.

Pimm, D. (1987). *Speaking mathematically: Communication in the mathematics classroom.* London: Routledge & Kegan Paul.

Pirie, S., & Schwarzenberger, R. (1988). Mathematical discussion and mathematical understanding. *Educational Studies in Mathematics, 19,* 459-470.

Poincaré, H. (1913). *The foundations of sciences.* New York: Science Press.

Poincaré, H. (1920). *Science and method.* New York: Dover.

Ponte, J. P. M. (1984). Functional reasoning and the interpretation of Cartesian graphs (Doctoral dissertation, University of Georgia, 1984). *Dissertation Abstracts International, 45*(6), 1675-A. (University Microfilms No. 8421144)

Ponte, J., & Abrantes, P. (1987). Teachers' views and attitudes about classroom computer use. In J. C. Bergeron, N. Herscovics, & C. Kieran (Eds.), *Proceedings of the 11th International Conference for the Psychology of Mathematics Education* (Vol. 2, pp. 71-77). Montréal, Canada: Université de Montréal.

Porteous, K. (1986). Children's appreciation of the significance of proof. In *Proceedings of the 10th International Conference for the Psychology of Mathematics Education* (Vol. 1, pp. 392-397). London: University of London, Institute of Education.

Putnam, R. T., Lesgold, S. B., Resnick, L. B., & Sterrett, S. G. (1987). Understanding sign change transformation. In J. C. Bergeron, N. Herscovics, & C. Kieran (Eds.), *Proceedings of the 11th International Conference for the Psychology of Mathematics Education* (Vol. 1, pp. 338-344). Montréal, Canada: Université de Montréal.

Pycior, H. M. (1984). Internalism, externalism and beyond: 19th century British algebra. *Historia Mathematica, 11,* 424-441.

Rasolofoniaina, I. (1984). Conditions d'apprentissage mathématique par la lecture. *Recherches en Didactique des Mathématiques, 5*(1), 5-42.

Razel, M., & Eylon, B. S. (1986). Developing visual language skills: The Agam program. *Journal of Visual Verbal Languaging, 6*(1), 49-54.

Resnick, L. B., & Nelson-Le Gall, S. (1987). Meaning construction in mathematical problem solving. In J. C. Bergeron, N. Herscovics, & C. Kieran (Eds.), *Proceedings of the 11th International Conference for the Psychology of Mathematics Education* (Vol. 3, pp. 215-220). Montréal, Canada: Université de Montréal.

Riley, M. S., Greeno, J. G., & Heller, J. I. (1983). Development of children's problem-solving ability in arithmetic. In H. Ginsburg (Ed.), *The development of mathematical thinking* (pp. 153-196). New York: Academic Press.

Rizutti, J., & Confrey, J. (1988). A construction of the concept of exponential function. In M. J. Behr, C. B. Lacampagne, & M. M. Wheeler (Eds.), *Proceedings of the 10th Annual Meeting of the North American Branch of the International Group for the Psychology of Mathematics Education* (pp. 260-267). DeKalb: Northern Illinois University.

Robert, A. (1982). L'acquisition de la notion de convergence des suites numériques dans l'enseignement supérieur. *Recherches en Didactique des Mathématiques, 3*(3), 305-341.

Robert, A., & Boschet, F. (1984). Acquisition des premiers concepts d'analyse sur R dans une section ordinaire de première année de DEUG [Entire issue]. *Cahier de Didactique, 7.* Paris: Institut de Recherche sur l'Enseignement des Mathématiques.

Rogalski, J. (1985). Acquisition of number-space relationships: Using educational and research programs. In L. Streefland (Ed.), *Proceedings of the Ninth International Conference for the Psychology of Mathematics Education* (Vol. 1, pp. 71-76). Utrecht, The Netherlands: State University of Utrecht, Subfaculty of Mathematics, OW & OC.

Rogalski, J., & Robert, A. (1988). Teaching and learning methods for problem solving: Some theoretical issues and psychological hypotheses. In A. Borbàs (Ed.), *Proceedings of the 12th International Conference for the Psychology of Mathematics Education* (Vol. 2, pp. 528-535). Veszprém, Hungary: Ferenc Genzwein, OOK.

Romberg, T. A. (1981). Towards a research consensus in some problem areas in the learning and teaching of mathematics. *Recherches en Didactique des Mathématiques, 2*(3), 347-362.

Romberg, T. A., & Stewart, D. M. (Eds.). (1987). *The monitoring of school mathematics: Background papers.* Madison: University of Wisconsin, Wisconsin Center for Education Research.

Rosch, E., & Mervis, C. B. (1975). Family resemblances: Studies in the internal structures of categories. *Cognitive Psychology, 7,* 578-605.

Rouchier, A. (1981). Problémes, procédures, programmes étudiés et réalisés par des enfants de CM2 utilisant un mini-ordinateur. *Revue Française de Pédagogie, 56,* 18-26.

Rouchier, A. (1987). La structure itérativo-répétitive comme champ conceptuel dans l'enseignement élémentaire et secondaire en mathématiques et informatique. In J. C. Bergeron, N. Herscovics, & C. Kieran (Eds.), *Proceedings of the 11th International Conference for the Psychology of Mathematics Education* (Vol. 3, pp. 76-83). Montréal, Canada: Université de Montréal.

Russell, B. (1919). *An introduction to mathematical philosophy.* London: Allen & Unwin.

Saada, E. H., & Brun, J. (1984). L'élaboration de formulations dans un jeu arithmétique. *Recherches en Didactique des Mathématiques, 2*(2), 215-231.

Samurçay, R. (1985). Learning programming: Constructing the concept of variable by beginning students. In L. Streefland (Ed.), *Proceedings of the Ninth International Conference for the Psychology of Mathematics Education* (Vol. 1, pp. 77-82).

Utrecht, The Netherlands: State University of Utrecht, Subfaculty of Mathematics, OW & OC.

Sastre, G., & Moreno, M. (1976). Représentation graphique de la quantité. *Bulletin de Psychologie de l'Université de Paris, 30,* 346-355.

Saxe, G. B. (1982). Culture and the development of numerical cognition: Studies among the Oksapmin of Papua New Guinea. In C. J. Brainerd (Ed.), *Progress in cognitive development research: Vol. 1. Children's logical and mathematical cognition* (pp. 157-176). New York: Spinger-Verlag.

Scally, S. P. (1986). A clinical investigation of the impact of a Logo learning environment on students' van Hiele levels of geometric understanding. In *Proceedings of the 10th International Conference for the Psychology of Mathematics Education* (Vol. 1, pp. 123-128). London: University of London, Institute of Education.

Scally, S. P. (1987). The effects of learning Logo on ninth grade students' understanding of geometric relations. In J. C. Bergeron, N. Herscovics, & C. Kieran (Eds.), *Proceedings of the 11th International Conference for the Psychology of Mathematics Education* (Vol. 2, pp. 46-52). Montréal, Canada: Université de Montréal.

Schaaf, W. L. (1930). Mathematics and world history. *Mathematics Teacher, 23,* 496-503.

Schaeffer, B., Eggleston, V., & Scott, J. L. (1974). Number development in young children. *Cognitive Psychology, 6,* 357-379.

Schoenfeld, A. H. (1982). Psychological factors affecting students' performance on geometry problems. In S. Wagner (Ed.), *Proceedings of the Fourth Annual Meeting of the North American Branch of the International Group for the Psychology of Mathematics Education* (pp. 168-174). Athens: University of Georgia.

Schoenfeld, A. H. (1985). *Mathematical problem solving.* New York: Academic Press.

Schoenfeld, A. H. (1986). On having and using geometric knowledge. In J. Hiebert (Ed.), *Conceptual and procedural knowledge: The case of mathematics* (pp. 225-264). Hillsdale, NJ: Erlbaum.

Schoenfeld, A. H. (1987). What's all the fuss about metacognition? In A. H. Schoenfeld (Ed.), *Cognitive science and mathematics education* (pp. 189-215). Hillsdale, NJ: Erlbaum.

Schubauer-Leoni, M. L., & Perret-Clermont, A. N. (1984). *Construction sociale d'écritures symboliques en deuxième primaire* (Interactions didactiques, No. 4). Geneva & Neuchâtel: Université de Genève, FASPE, & Université de Neuchâtel, Séminaire de psychologie.

Schwartz, J. L., & Yerushalmy, M. (1987). The Geometric Supposer: An intellectual prosthesis for making conjectures. *College Mathematics Journal, 18,* 58-65.

Schwartz, J. L., Yerushalmy, M., & Education Development Center. (1985). *The Geometric Supposer* [computer program]. Pleasantville, NY: Sunburst Communications.

Schwarz, B. (1989). *The use of a microworld to improve the concept image of a function: The triple representation model curriculum.* Unpublished doctoral dissertation, Weizmann Institute of Science, Rehovot, Israel.

Schwarz, B., & Bruckheimer, M. (1988). Representations and analogies. In A. Borbàs (Ed.), *Proceedings of the 12th International Conference for the Psychology of Mathematics Education* (Vol. 2, pp. 552-559). Veszprém, Hungary: Ferenc Genzwein, OOK.

Schwarz, B., & Dreyfus, T. (1989). Transfer between functional representations: A computational model. In G. Vergnaud, J. Rogalski, & M. Artigue (Eds.), *Proceedings of the 13th International Conference for the Psychology of Mathematics Education* (Vol. 3, pp. 143-150). Paris: Laboratoire PSYDEE.

Semadeni, Z., & Puchalska, E. (1987). Verbal problems with missing, surplus or contradictory data. *Séminaire de Didactique des Mathématiques et de l'Informatique* (LSD-IMAG, Institut Fourier, Grenoble, France, année 1986-1987 et 1987-1988), pp. 139-151.

Sfard, A. (1987). Two conceptions of mathematical notions: Operational and structural. In J. C. Bergeron, N. Herscovics, & C. Kieran (Eds.), *Proceedings of the 11th International Conference for the Psychology of Mathematics Education* (Vol. 3, pp. 162-169). Montréal, Canada: Université de Montréal.

Sfard, A. (1988). Operational vs. structural method of teaching mathematics--Case study. In A. Borbàs (Ed.), *Proceedings of the 12th International Conference for the Psychology of Mathematics Education* (Vol. 2, pp. 560-567). Veszprém, Hungary: Ferenc Genzwein, OOK.

Sfard, A. (1989). Transition from operational to structural conception: The notion of function revisited. In G. Vergnaud, J. Rogalski, & M. Artigue (Eds.), *Proceedings of the 13th International Conference for the Psychology of Mathematics Education* (Vol. 3, pp. 151-158). Paris: Laboratoire PSYDEE.

Sharan, S., Hare, P., Webb, C. D., & Hertz-Lazarowitz, R. (1980). *Cooperation in education.* Provo, UT: Brigham Young University Press.

Shavelson, R. L., Winckler, J. D., & Stasz, C. (1985). Patterns of microcomputer use in teaching mathematics and science. *Journal of Educational Computing Research, 1,* 395-413.

Shelton, M. (1985). Geometry, spatial development and computers: Young children and triangle concept development. In S. K. Damarin & M. Shelton (Eds.), *Proceedings of the Seventh Annual Meeting of the North American Branch of the International Group for the Psychology of Mathematics Education* (pp. 256-261). Columbus: Ohio State University.

Shumway, R. (Ed.). (1980). *Research in mathematics education.* Washington, DC: National Council of Teachers of Mathematics.

Siegler, R. S., & Shrager, J. (1984). Strategy choices in addition and subtraction: How do children know what to do? In C. Sophian (Ed.), *Origins of cognitive skills* (pp. 229-294). Hillsdale, NJ: Erlbaum.

Sierpinska, A. (1985). Obstacles épistémologiques relatifs à la notion de limite. *Recherches en Didactique des Mathématiques, 6*(1), 5-67.

Silver, E. A. (Ed.). (1985). *Teaching and learning mathematical problem solving: Multiple research perspectives.* Hillsdale, NJ: Erlbaum.

Simon, H. A. (1981). *The sciences of the artificial.* Cambridge, MA: MIT Press.

Sinclair, A., Siegriest, F., & Sinclair, H. (1983). Young children's ideas about the written number system. In D. Rogers & J. Sloboda (Eds.), *The acquisition of symbolic skills* (pp. 535-542). New York: Plenum.

Sinclair, H. (1987). Constructivism and the psychology of mathematics. In J. C. Bergeron, N. Herscovics, & C. Kieran (Eds.), *Proceedings of the 11th International Conference for the Psychology of Mathematics Education* (Vol. 1, pp. 28-41). Montréal, Canada: Université de Montréal.

Skemp, R. (1971). *The psychology of learning mathematics.* Harmondsworth, UK: Penguin.

Skemp, R. R. (1979). *Intelligence, learning, and action.* Chichester, U.K.: Wiley.

Skemp, R. (Ed.). (1982). Understanding the symbolism of mathematics [Special issue]. *Visible Language, 16*(3).

Sleeman, D. H., & Brown, J. S. (Eds.). (1982). *Intelligent tutoring systems.* London: Academic Press.

Soloway, E., Lochhead, J., & Clement, J. (1982). Does computer programming enhance problem solving ability? Some positive evidence on algebra word problems. In R. J. Seidel, R. E. Anderson, & B. Hunter (Eds.), *Computer literacy* (pp. 171-185). New York: Academic Press.

Southwell, B. (1985). The use of calculators to develop reasoning processes and achievement in problem solving. In L. Streefland (Ed.), *Proceedings of the Ninth International Conference for the Psychology of Mathematics Education* (Vol. 1, pp. 90-100). Utrecht, The Netherlands: State University of Utrecht, Subfaculty of Mathematics, OW & OC.

Sowder, L. K. (1980). Concept and principle learning. In R. J. Shumway (Ed.), *Research in mathematics education* (pp. 244-285). Reston, VA: National Council of Teachers of Mathematics.

Spanos, G., Rhodes, N. C., Dale, T. C., & Crandall, J. (1988). Linguistic features of mathematics problem solving: Insights and applications. In R. R. Cocking & J. P. Mestre (Eds.), *Linguistic and cultural influences on learning mathematics* (pp. 221-240). Hillsdale, NJ: Erlbaum.

Steen, L. A. (Ed.). (1987). *Calculus for a new century: A pump, not a filter* (MAA Notes No. 8). Washington, DC: Mathematical Association of America.

Steffe, L. P., von Glasersfeld, E., Richards, J., & Cobb, P. (1983). *Children's counting types: Philosophy, theory, and application.* New York: Praeger.

Sullivan, H. S. (1955). *Conceptions of modern psychiatry.* Southampton, PA: Burgundy Press.

Sutherland, R. (1987a). A study of the use and understanding of algebra related concepts within a Logo environment. In J. C. Bergeron, N. Herscovics, & C. Kieran (Eds.), *Proceedings of the 11th International Conference for the Psychology of Mathematics Education* (Vol. 1, pp. 241-247). Montréal, Canada: Université de Montréal.

Sutherland, R. (1987b). What are the links between variable in Logo and variable in algebra. *Recherches en Didactique des Mathématiques, 8*(1-2), 103-130.

Sutherland, R., & Hoyles, C. (1986). Logo as a context for learning about variable. In *Proceedings of the 10th International Conference for the Psychology of Mathematics Education* (Vol. 1, pp. 301-306). London: University of London, Institute of Education.

Tall, D. (1985). Using computer graphics programs as generic organisers for the concept image of differentiation. In L. Streefland (Ed.), *Proceedings of the Ninth International Conference for the Psychology of Mathematics Education* (Vol. 1, pp. 105-110). Utrecht, The Netherlands: State University of Utrecht, Subfaculty of Mathematics, OW & OC.

Tall, D. (1986). *Building and testing a cognitive approach to the calculus using interactive computer graphics.* Doctoral dissertation, University of Warwick, Mathematics Education Research Centre, Warwick, UK.

Tall, D. (1987a). Algebra in a computer environment. In J. C. Bergeron, N. Herscovics, & C. Kieran (Eds.), *Proceedings of the 11th International Conference for the Psychology of Mathematics Education* (Vol. 1, pp. 262-274). Montréal, Canada: Université de Montréal.

Tall, D. (1987b). Constructing the concept image of a tangent. In J. C. Bergeron, N. Herscovics, & C. Kieran (Eds.), *Proceedings of the 11th International Conference for the Psychology of Mathematics Education* (Vol. 3, pp. 69-75). Montréal, Canada: Université de Montréal.

Tall, D., & Vinner, S. (1981). Concept images and concept definition in mathematics with particular reference to limits and continuity. *Educational Studies in Mathematics, 12,* 151-169.

Taloumis, T. (1975). The relationship of area conservation to area measurement as affected by sequence of presentation of Piagetian area tasks to boys and girls in grades one to three. *Journal for Research in Mathematics Education, 6,* 233-242.

Thom, R. (1972). Modern mathematics: Does it exist? In A. G. Howson (Ed.), *Developments in mathematical education: Proceedings of the Second International*

Congress on Mathematical Education (pp. 194-209).   Cambridge: Cambridge University Press.

Thomas, H. L. (1975). The concept of function. In M. F. Rosskopf (Ed.), *Children's mathematical concepts: Six Piagetian studies in mathematics education* (pp. 145-172). New York: Teachers College Press.

Thomas, M., & Tall, D. (1986). The value of the computer in learning algebra concepts. In *Proceedings of the 10th International Conference for the Psychology of Mathematics Education* (Vol. 1, pp. 313-318). London: University of London, Institute of Education.

Thompson, A. (1982). *Teachers' conceptions of mathematics and mathematics teaching: Three case studies.* Unpublished doctoral dissertation, University of Georgia, Athens.

Thompson, A. (1984). The relationship of teachers' conceptions of mathematics and mathematics teaching to instructional practice. *Educational Studies in Mathematics, 15,* 105-127.

Thompson, P. W. (1985). Experience, problem solving, and learning mathematics: Considerations in developing mathematics curricula. In E. A. Silver (Ed.), *Teaching and learning mathematical problem solving: Multiple research perspectives* (pp. 189-236). Hillsdale, NJ: Erlbaum.

Thompson, P. W., & Dreyfus, T. (1988). Integers as transformations. *Journal for Research in Mathematics Education, 19,* 115-133.

Thompson, P. W., & Thompson, A. G. (1987). Computer presentations of structure in algebra. In J. C. Bergeron, N. Herscovics, & C. Kieran (Eds.), *Proceedings of the 11th International Conference for the Psychology of Mathematics Education* (Vol. 1, pp. 248-254). Montréal, Canada: Université de Montréal.

Thorndike, E. L. (1922). *The psychology of arithmetic.* New York: MacMillan.

Tirosh, D., Fischbein, E., & Dor, E. (1985). The teaching of infinity. In L. Streefland (Ed.), *Proceedings of the Ninth International Conference for the Psychology of Mathematics Education* (Vol. 1, pp. 501-506). Utrecht, The Netherlands: State University of Utrecht, Subfaculty of Mathematics, OW & OC.

Travers, K. J. (Director). (1986). *Second international mathematics study: Detailed report for the United States.* Champaign, IL: Stipes.

Usiskin, Z. (1982). *Van Hiele levels and achievement in secondary school geometry* (CDASSG Project Report). Chicago: University of Chicago, School of Education.

Van der Waerden, B. L. (1983). *Algebra and geometry in ancient civilisations.* Berlin: Springer-Verlag.

Van Hiele, P. M. (1987). *A method to facilitate the finding of levels of thinking in geometry by using the levels in arithmetic.* Paper presented at the Conference on Learning and Teaching Geometry, Syracuse University, Syracuse, NY.

Van Hiele, P. M., & van Hiele-Geldof, D. (1958). A method of initiation into geometry. In H. Freudenthal (Ed.), *Report on methods of initiation into geometry.* Groningen: Walters.

Vergnaud, G. (1981a). *L'enfant, la mathématique et la réalité.* Bern: Peter Lang.

Vergnaud, G. (1981b). Quelques orientations théoriques et méthodologiques des recherches françaises en didactique des mathématiques. *Recherches en Didactique des Mathématiques, 2*(2), 215-231.

Vergnaud, G. (1982a). A classification of cognitive tasks and operations of thought involved in addition and subtraction problems. In T. P. Carpenter, J. M. Moser, & T. A. Romberg (Eds.), *Addition and subtraction: A cognitive perspective* (pp. 39-59). Hillsdale, NJ: Erlbaum.

Vergnaud, G. (1982b). Cognitive and developmental psychology and research in mathematics education: Some theoretical and methodological issues. *For the Learning of Mathematics, 3*(2), 31-41.

Vergnaud, G. (1984). Understanding mathematics at the secondary-school level. In A. Bell, B. Low, & J. Kilpatrick (Eds.), *Theory, research and practice in mathematical education* (Report of ICME 5 Working Group on Research in Mathematics Education, pp. 27-35). Nottingham, UK: Shell Centre for Mathematical Education.

Vergnaud, G. (1987). About constructivism. In J. C. Bergeron, N. Herscovics, & C. Kieran (Eds.), *Proceedings of the 11th International Conference for the Psychology of Mathematics Education* (Vol. 1, pp. 42-54). Montréal, Canada: Université de Montréal.

Vergnaud, G. (1988). Long terme et court terme dans l'apprentissage de l'algèbre. In C. Laborde (Ed.), *Actes du Premier Colloque Franco-Allemand de Didactique des Mathématiques et de l'Informatique* (pp. 189-199). Grenoble: La Pensée Sauvage.

Vergnaud, G., & Errecalde, P. (1980). Some steps in the understanding and the use of scales and axis by 10-13 year-old students. In R. Karplus (Ed.), *Proceedings of the Fourth International Conference for the Psychology of Mathematics Education* (pp. 285-291). Berkeley: University of California.

Verschaffel, L., & De Corte, E. (in press). Do non-semantic factors also influence the solution process of addition and subtraction word problems? In H. Mandl, E. De Corte, S. N. Bennett, & H. F. Friedrich (Eds), *Learning and instruction: European research in an international context: Vol. 3. Analysis of teaching, complex skills, and complex knowledge domains.* Oxford: Pergamon Press.

Vinner, S. (1981). The nature of geometrical objects as conceived by teachers and prospective teachers. In C. Comiti & G. Vergnaud (Eds.), *Proceedings of the Fifth International Conference for the Psychology of Mathematics Education* (pp. 375-380). Grenoble, France: Institut IMAG.

Vinner, S. (1982). Conflicts between definitions and intuitions--The case of the tangent. In A. Vermandel (Ed.), *Proceedings of the Sixth International Conference for the Psychology of Mathematics Education* (pp. 24-28).   Antwerp, Belgium: Universitaire Instelling Antwerpen.

Vinner, S. (1983). Concept definition, concept image and the notion of function. *International Journal of Mathematical Education in Science and Technology, 14,* 239-305.

Vinner, S. (1987). Continuous functions--Images and reasoning in college students. In J. C. Bergeron, N. Herscovics, & C. Kieran (Eds.), *Proceedings of the 11th International Conference for the Psychology of Mathematics Education* (Vol. 3, pp. 177-183). Montréal, Canada: Université de Montréal.

Vinner, S., & Dreyfus, T. (1989). Images and definitions for the concept of function. *Journal for Research in Mathematics Education, 20,* 356-366.

Vinner, S., & Hershkowitz, R. (1980). Concept images and common cognitive paths in the development of some simple geometrical concepts. In R. Karplus (Ed.), *Proceedings of the Fourth International Conference for the Psychology of Mathematics Education* (pp. 177-184). Berkeley: University of California.

Vinner, S., & Hershkowitz, R.   (1983).   On concept formation in geometry. *Zentralblatt für Didaktik der Mathematik, 15,* 20-25.

Von Glasersfeld, E. (1983). Learning as a constructive activity. In J. C. Bergeron & N. Herscovics (Eds.), *Proceedings of the Fifth Annual Meeting of the North American Chapter of the International Group for the Psychology of Mathematics Education* (Vol. 1, pp. 41-69). Montréal, Canada: Université de Montréal.

Von Glasersfeld, E. (1984). Introduction. In P. Watzlawick (Ed.), *The invented reality* (pp. 13-15). New York: Norton.

Von Glasersfeld, E. (1986). Steps in the construction of "others" and "reality": A study in self-regulation. In R. Trappe (Ed.), *Power, autonomy, utopia* (pp. 107-116). New York: Plenum.

Vygotsky, L. S. (1962).   *Thought and language.*   Cambridge, MA: MIT Press. (Original work published 1934.)

Wagner, S., Rachlin, S. L., & Jensen, R. J. (1984). *Algebra Learning Project: Final report.* Athens: University of Georgia, Department of Mathematics Education.

Webb, N.   (1982).   Group composition, group interaction and achievement in cooperative small groups. *Journal of Educational Psychology, 74,* 475-484.

Wellman, H. (1985). The origins of metacognition. In D. L. Forrest-Pressley, G. E. MacKinnon, & T. G. Waller (Eds.). *Metacognition, cognition, and human performance* (Vol. 1, pp. 1-31). London: Academic Press.

Wenger, E. (1987). *Artificial intelligence and tutoring systems.* Los Altos, CA: Morgan Kaufmann.

Wheeler, D. (1987). The world of mathematics: Dream, myth or reality? In J. C. Bergeron, N. Herscovics, & C. Kieran (Eds.), *Proceedings of the 11th International Conference for the Psychology of Mathematics Education* (Vol. 1, pp. 55-66). Montréal, Canada: Université de Montréal.

Wheeler, D. (1989). Contexts for research on the teaching and learning of algebra. In S. Wagner & C. Kieran (Eds.), *Research agenda for mathematics education: Vol. 4. Research issues in the learning and teaching of algebra* (pp. 278-287). Hillsdale, NJ: Erlbaum.

Wheeler, M. M., & Martin, W. G. (1988). Explicit knowledge of infinity. In M. J. Behr, C. B. Lacampagne, & M. M. Wheeler (Eds.), *Proceedings of the 10th Annual Meeting of the North American Branch of the International Group for the Psychology of Mathematics Education* (pp. 312-318). DeKalb: Northern Illinois University.

Whitman, B. (1976). Intuitive equation solving skills and the effects on them of formal techniques of equation solving (Doctoral dissertation, Florida State University, 1975). *Dissertation Abstracts International, 36,* 5180A. (University Microfilms No. 76-2720)

Wilson, P. S. (1983). Use of negative instances in identifying geometric features. In J. C. Bergeron & N. Herscovics (Ed.), *Proceedings of the Fifth Annual Meeting of the North American Branch of the International Group for the Psychology of Mathematics Education* (Vol. 1, pp. 326-332). Montréal, Canada: Université de Montréal.

Wilson, P. S. (1986). The relationship between children's definitions of rectangles and their choice of examples. In G. Lappan & R. Even (Eds.), *Proceedings of the Eighth Annual Meeting of the North American Branch of the International Group for the Psychology of Mathematics Education* (pp. 158-162). East Lansing: Michigan State University.

Wirszup, I. (1976). Breakthroughs in the psychology of learning and teaching geometry. In J. L. Martin (Ed.), *Space and geometry* (pp. 75-98). Columbus, OH: ERIC/SMEAC.

Yackel, E. (1987). A year in the life of a second grade class: A small group perspective. In J. C. Bergeron, N. Herscovics, & C. Kieran (Eds.), *Proceedings of the 11th International Conference for the Psychology of Mathematics Education* (Vol. 3, pp. 208-214). Montréal, Canada: Université de Montréal.

Yerushalmy, M. (1988). *Effects of graphic feedback on the ability to transform algebraic expressions when using computers* (Interim report to the Spencer Fellowship Program of the National Academy of Education). Haifa, Israel: University of Haifa.

Yerushalmy, M., & Chazan, D. (in press). Overcoming visual obstacles with the aid of the Supposer. In T. Butler & J. L. Schwartz (Eds.), *The Geometric Supposer reader.* Hillsdale, NJ: Erlbaum.

Yerushalmy, M., Chazan, D., Gordon, M., & Houde, R. (1986). Micro-computer centered plane geometry teaching. In G. Lappan & R. Even (Eds.), *Proceedings of the Eighth Annual Meeting of the North American Branch of the International Group for the Psychology of Mathematics Education* (pp. 184-189). East Lansing: Michigan State University.

Young, A. W., & McPherson, J. (1976). Ways of making number judgments and children's understanding of quantity relations. *British Journal of Educational Psychology, 46,* 328-332.

Zehavi, N. (1988). The effect of mathematics software on shaping students' intuitions. *Journal for Educational Computing Research, 4,* 391-401.

Zykova, V. I. (1969). Operating with concepts when solving geometry problems. In J. Kilpatrick & I. Wirszup (Eds.), *Soviet studies in the psychology of learning and teaching mathematics: Vol. 1. The learning of mathematical concepts* (pp. 93-148). Stanford, CA: School Mathematics Study Group.